编委会

主　编

俞汉青

编　委

（以姓氏拼音排序）

陈洁洁　刘　畅　刘武军

刘贤伟　卢　姝　吕振婷

裴丹妮　盛国平　孙　敏

汪雯岚　王楚亚　王龙飞

王维康　王允坤　徐　娟

俞汉青　虞盛松　院士杰

翟林峰　张爱勇　张　锋

"十四五"国家重点出版物出版规划重大工程

二氧化钛纳米单晶的电化学功能及其在水处理中的应用

Electrochemical Functions of TiO$_2$ Single Crystals and Their Applications in Water Treatment

俞汉青 著
刘 畅

中国科学技术大学出版社

内容简介

电化学技术在水污染控制领域具有重要的研究和应用价值,其在低偏压条件下能快速、低能耗地去除污染物。但是该技术目前缺乏廉价、高效的电催化剂,且电极易于污染,严重限制了实际应用。二氧化钛(TiO_2)是一种良好的光催化材料,其结构稳定、廉价、安全,具有较高的电催化功能开发和实际应用价值,但是其电导率和电化学活性较低,难以直接作为电催化剂使用。针对这些问题,本书以 TiO_2 电极材料为核心,较为全面地讨论了其电化学功能拓展的前沿方法,包括单晶结构设计、晶面工程和缺陷工程的调控手段等,涵盖了污染物电催化阳极氧化和阴极还原的基础理论知识,并对 TiO_2 光化学电极防污、污染物检测与电化学储能等重要水处理工艺与新功能应用成果进行了详细介绍。

图书在版编目(CIP)数据

二氧化钛纳米单晶的电化学功能及其在水处理中的应用/俞汉青,刘畅著. —合肥:中国科学技术大学出版社,2022.3
(污染控制理论与应用前沿丛书/俞汉青主编)
国家出版基金项目
"十四五"国家重点出版物出版规划重大工程
ISBN 978-7-312-05391-7

Ⅰ.二… Ⅱ.①俞… ②刘… Ⅲ.二氧化钛—单晶—纳米材料—应用—废水处理 Ⅳ.X703

中国版本图书馆 CIP 数据核字(2022)第 029813 号

二氧化钛纳米单晶的电化学功能及其在水处理中的应用
ERYANGHUATAI NAMI DANJING DE DIANHUAXUE GONGNENG
JI QI ZAI SHUI CHULI ZHONG DE YINGYONG

出版	中国科学技术大学出版社 安徽省合肥市金寨路96号,230026 http://www.press.ustc.edu.cn https://zgkxjsdxcbs.tmall.com
印刷	安徽联众印刷有限公司
发行	中国科学技术大学出版社
开本	787 mm×1092 mm 1/16
印张	16
字数	305 千
版次	2022 年 3 月第 1 版
印次	2022 年 3 月第 1 次印刷
定价	100.00 元

总　序

建设生态文明是关系人民福祉、关乎民族未来的长远大计,在党的十八大以来被提升到突出的战略地位。2017年10月,党的十九大报告明确提出"污染防治"是生态文明建设的重要战略部署,是我国决胜全面建成小康社会的三大攻坚战之一。2018年,国务院政府工作报告进一步强调要打好"污染防治攻坚战",确保生态环境质量总体改善。这都显示出党和国家推动我国生态环境保护水平同全面建成小康社会目标相适应的决心。

当前,我国环境污染状况有所缓解,但总体形势仍然严峻,已严重制约了我国经济社会的持续健康发展。发展以资源回收利用为导向的污染控制新理论与新技术,是进一步推动污染物高效、低成本、稳定去除的发展方向,已成为国家重大战略需求和国际重要学术前沿。

为了配合国家对生态文明建设、"污染防治攻坚战"的一系列重大布局,抢占污染控制领域国际学术前沿制高点,加快传播与普及生态环境污染控制的前沿科学研究成果,促进相关领域人才培养,推动科技进步及成果转化,我们组织一批来自多个"双一流"大学、活跃在我国环境科学与工程前沿领域、有影响力的科学家共同撰写"污染控制理论与应用前沿丛书"。

本丛书是作者团队承担的国家重大重点科研项目(国家重大科技专项、国家863计划、国家自然科学基金)和获得的重大科技成果奖励(2014年国家自然科学奖二等奖、2020年国家科学技术进步奖二等奖)的系统总结,是作者团队攻读博士学位期间取得的重要的前沿学术成果(全国百篇优秀博士论文、中科院优秀博士论文等)的系统凝练,是一套系统反映污染控制基础科学理论与前沿高新技术研究成果的系列图书。本丛书围绕我国环境领域的污染物生化控制、转化机理、无害化处置、资源回收利用等亟须解决的一些重大科学问题与技术问题,将物理学、化学、生物学、材料学等学科的最新理论成果以及前沿高新技术应用到污染控制过程中,总结了我国目前在污染控制领域(特别是废水和固废领域)的重要研究进展,探索、建立并发展了常温空气阴极燃料电池、纳米材料、新兴生物电化学系统、新型膜生物反应器、水体污染物的化学及生物转化,以及固体废弃物污染控制与清洁转化等方面的前沿理论与技术,形成了具有广阔应用前景的新理论和新方法,为污染控制与治理提供了理论基础和科学依据。

"污染控制理论与应用前沿丛书"是服务国家重大战略需求、推动生态文明建设、打赢"污染防治攻坚战"的一套丛书。其出版将有利于促进最前沿的科研成果得到及时的传播和应用,有利于促进污染治理人才和高水平创新团队的培养,有利于推动我国环境污染控制和治理相关领域的发展和国际竞争力的提升;同时为环境污染控制与治理实践提供新思路、新技术、新材料,也可以为政府环境决策、强化环境管理、履行国际环境公约等提供科学依据和技术支撑,在保障生态环境安全、实施生态文明建设、打赢"污染防治攻坚战"中起到不可替代的作用。

编委会

2021 年 10 月

前 言

电化学技术在水污染控制领域具有重要的研究和应用价值,其在低偏压条件下能快速、低能耗地去除污染物。但是目前该技术缺乏廉价、高效的电催化剂,且电极易于污染,严重限制了实际应用。二氧化钛(TiO_2)是一种良好的光催化材料,结构稳定、廉价、安全,其中锐钛矿晶型因 Ti、O 原子排布疏松、位错缺陷多、催化活性强,而具有较高的电催化功能开发和实际应用价值,但是其电导率和电化学活性较低,难以直接作为电催化剂使用。针对这些问题,本书以 TiO_2 电极材料为核心,着重探索、讨论其电化学功能拓展的前沿方法,并对 TiO_2 光化学电极防污、污染物检测与电化学储能等重要水处理工艺与新功能应用成果进行了详细介绍。本书涵盖的主要前沿进展如下:

(1) TiO_2 单晶的电催化氧化新机制的揭示

针对 TiO_2 电化学活性较低及阳极氧化机理不明确的问题现状,作者团队以 TiO_2 为本征材料,采用晶面调控策略合成了具有高能{001}极性晶面暴露的单晶电极材料。选取 5 种含吸电子或供电子基团的对位取代酚作为目标污染物,结合密度泛函理论计算和电催化表征实验,建立了污染物电催化氧化效率与对位取代基电学性质以及电极表面富集浓度之间的数学关系,揭示了 TiO_2 单晶的电催化氧化机制。研究发现,酚类污染物的降解主要通过吸附态·OH 氧化和直接电子转移完成。这些发现为 TiO_2 电极材料的制备、改性和污染物降解应用提供了新的思路。这些内容我们将在第 2 章进行详细阐述。

(2) 电催化阳极抗污染的新策略的发展

为解决污染物低压电化学转化中的电极污染问题,作者团队借助高能{001}晶面暴露 TiO_2 单晶光、电催化双重优异的属性,原位构建了紫外光辅助电催化氧化的耦合体系,并发展了阳极抗污染新策略。以环境内分泌干扰物双酚 A 和垃圾渗滤液为处理对象,在低外加偏压和紫外光照射条件下,研究了光辅助电催化体系对污染物和实际废水的处理效能。通过污染物循环降解实验、光化学表征、电化学表征、自由基捕获和表面化学分析,揭示了双功能 TiO_2 单晶介导的光辅助电催化体系的氧化效果、抗污染特性和作用机制。研究发现,低压电解产生的阳极污染聚合物可以通过游离态·OH 介导的光化学氧化途径进行有效去除,在循环降解过程中电极表面始终保持洁净,可以实现双酚 A 和垃圾渗滤液的连续高效处理。该工作构建了一种高效、节能的光电耦合催化水处理新体系。这些内容我们将在第 3 章进行详细阐述。

(3) 缺陷型电催化活性位点的保护新方法的建立

为克服缺陷型金属氧化物电极材料催化活性位点稳定性差的弊端,作者团队借助缺陷型 TiO_{2-x} 单晶完全不同的光、电催化激发路径,构建了可见光耦合电催化氧化新体系,并发展了保护缺陷活性位点的新方法。以双酚 A 和工业废水为处理对象,研究了 TiO_{2-x} 单晶光辅助电催化体系的氧化效能、活性位点保护和作用机制。结果表明,TiO_{2-x} 表面和亚表面的缺陷型催化活性位点可以通过可见光非带隙激发路径得到保护,并在光化学作用下具备阳极抗污染特性,因此具有良好的催化活性和稳定性。该工作进一步拓宽了 TiO_2 在电化学水处理工艺中的应用前景。这些内容我们将在第 4 章进行详细阐述。

(4) TiO_{2-x} 单晶电催化氧还原新功能的拓展

针对电催化氧还原应用中贵金属 Pt 用量大、成本高以及稳定性差的问题,作者团队以 TiO_2 为出发材料,借助晶体缺陷调控策略,制备出表面和亚表面富含氧空位的缺陷型单晶电极材料,证实其具有优异的氧还原反应活性、稳定性和抗甲醇氧化性;同时结合密度泛函理论计算,揭示了面导向氧吸附构型、缺陷中心氧还原反应机理及两电子还原机制。研究发现,TiO_2 电导率和氧还原活性的提高,得益于单晶结构、高暴露{001}晶面、Ti^{3+}/氧空位的设计调控及与 rGO 的有效复合。该工作开发了 TiO_2 的电催化新功能,为其在环境、能源领域的拓展应用提供了新思路。这些内容我们将在第 5 章进行详细阐述。

(5) TiO_{2-x} 单晶阴极电催化还原功能的开发

针对污染物电催化还原处理中优异阴极材料缺失的难题,作者团队以难以氧化处理的硝基苯作为目标污染物,研究了缺陷型 TiO_{2-x} 单晶对硝基苯的电催化还原效率。通过硝基苯降解实验、电化学表征、自由基捕获和还原产物分析,揭示了缺陷型 TiO_{2-x} 单晶对硝基苯的电催化还原效能和机制。结果表明,晶体形状、暴露晶面和化学计量学氧含量的局部电子结构精细调控,使 TiO_2 成为一种良好的非贵金属阴极催化剂,可以实现硝基苯的高效、稳定还原,进而拓展了其在环境领域的应用范围。这些内容我们将在第 6 章进行详细阐述。

(6) 分子印迹功能化 TiO_2 单晶的电化学检测新方法的构建

针对电化学直接氧化法检测 BPA 所面临的响应低、选择性差等问题,作者团队通过在材料水热合成过程中预先加入 BPA,制备出具有特定无机框架和分子识别能力的 {001}-TiO_2 单晶。研究发现,经过 BPA 预处理后,TiO_2 单晶对痕量 BPA 的电化学检测展现出较佳的灵敏度和选择性,且检测区分能力显著增强,具有良好的抗干扰能力和循环稳定性,能够胜任自来水、地表水和生活污水等实际环境样品的检测测试。该工作拓宽了 TiO_2 的电催化应用范围,为发展高活性、高稳定性、低成本且环境友好的电化学传感电极材料提供了理论支撑。这些内容我们将在第 7 章进行详细阐述。

本书的出版得到了国家出版基金和国家自然科学基金的大力支持,在此致以衷心的感谢!

由于编者水平有限,书中难免有疏漏和不妥之处,敬请读者批评指正。

目 录

总序 —— i

前言 —— iii

第 1 章
TiO$_2$ 的电化学水处理潜能

1.1 电化学水处理技术的优势与挑战 —— 3

1.2 TiO$_2$ 在环境领域中的研究现状 —— 9

1.3 TiO$_2$ 电化学水处理的应用挑战与解决策略 —— 11

第 2 章
TiO$_2$ 单晶电催化降解酚类污染物

2.1 概述 —— 29

2.2 TiO$_2$ 单晶电极的合成与电催化体系设计 —— 30

2.3 TiO$_2$ 单晶电催化降解对位取代酚的效能与机理分析 —— 37

第 3 章
紫外光辅助 TiO$_2$ 单晶电催化降解污染物

3.1 概述 —— 69

3.2 TiO$_2$ 单晶的紫外光-电性能与光电耦合催化体系设计 —— 70

3.3 紫外光辅助 TiO$_2$ 单晶电催化降解双酚 A 的效能与机理分析 —— 74

第 4 章
可见光辅助 TiO$_{2-x}$ 单晶电催化降解污染物

4.1 概述 —— 105

4.2 TiO$_{2-x}$ 单晶的可见光-电性能与光电耦合催化体系设计 —— 106

4.3 可见光辅助 TiO$_{2-x}$ 单晶电催化降解双酚 A 的效能与机理分析 —— 110

第 5 章
TiO$_{2-x}$ 单晶的电催化氧还原新功能

5.1 概述 —— 159

5.2 TiO$_{2-x}$ 单晶的合成新方法与氧还原体系设计 —— 160

5.3 TiO$_{2-x}$ 单晶的电催化氧还原性能与机理分析 —— 163

第 6 章
TiO$_{2-x}$ 单晶电催化还原降解硝基苯

6.1 概述 —— 187

6.2 TiO$_{2-x}$ 单晶的阴极性能表征与电催化还原体系设计 —— 188

6.3 TiO$_{2-x}$ 单晶电催化还原硝基苯的效能与机理分析 —— 190

第 7 章
分子印迹功能化 TiO$_2$ 单晶电化学检测双酚 A

7.1 概述 —— 215

7.2 分子印迹功能化 MI-TiO$_2$ 单晶的电化学传感性能表征 —— 216

7.3 MI-TiO$_2$ 单晶电化学检测双酚 A 的效能与机理分析 —— 220

第 1 章

TiO$_2$ 的电化学水处理潜能

1.1 电化学水处理技术的优势与挑战

随着社会经济的发展、城市化和工业化进程的加快,水污染问题日益严峻,开发新型高效的水处理技术迫在眉睫。由于工业废水同时含有无机物和有机物,成分非常复杂,因而对资源难以进行有效回收。对于有机废水的处理,生物氧化法最为廉价;但高毒性和难以生物降解物质的存在,使生物氧化法的应用受到严重限制。除生物氧化法之外,目前常用的水处理技术还有物理法(过滤、吸附和混凝等)、化学氧化法(使用氯、臭氧和过氧化氢等进行氧化)和高级氧化法(芬顿反应、臭氧+紫外光辐照和光催化等)。然而,这些方法都存在技术缺陷,例如,过滤吸附处理无法完全达到排放标准[1],混凝浮选会产生大量污泥,化学氧化驱动力有限且需要运输和储存不稳定的危险试剂,且部分高级氧化法往往需要较高的投资成本,因而均不是理想的水处理技术。

在此背景下,电化学技术为工业废水的处理提供了一种有效的解决方案(图1.1),

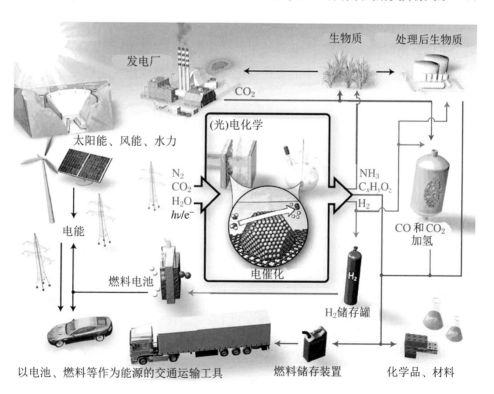

图 1.1 电化学技术的应用领域[2]

因为电子是一种多功能、高效、经济、易于控制且清洁的试剂[3]。近几十年来,电化学工艺装置一直广泛应用于金属离子去除回收、含铬废液处理、废气脱硫和海水淡化脱盐等工业生产线,而在处理有机污染物方面的功能作用"相对较小"[4-7];如今随着科学研究的不断深入,电极材料的催化活性和稳定性都得到了有效提高,反应器的几何结构逐步得到优化,使电化学技术的发展突飞猛进,前景愈加广阔,足以胜任各类水体的消毒和净化[8-9]。

1.1.1
电化学氧化过程分类

在电化学氧化体系中,污染物的氧化可分为直接氧化和间接氧化两种形式。直接氧化是指阳极材料表面发生与污染物的直接电子转移过程,没有其他物质参与反应;间接氧化则是有机污染物与电化学反应生成的活性物质发生电子转移的过程,此过程中污染物不直接接触阳极表面,由活性物质充当污染物与阳极材料发生电子交换的媒介。间接氧化可以是可逆的,也可以是不可逆的;氧化还原反应物可以在阳极生成,也可以在阴极生成。电化学氧化体系的反应模式,取决于电极材料的性质、结构、实验条件和介质组成。

以下对电化学反应体系中的直接氧化和间接氧化过程作一些简要介绍。

1.1.1.1　直接氧化

在直接氧化过程中,污染物首先吸附在阳极表面,随后直接发生氧化反应,不涉及除电子之外的任何物质,是一种"清洁技术":

$$R_{ads} - ze^- \rightarrow P_{ads} \tag{1.1}$$

在电化学矿化过程中,氧原子通过电驱动力直接转移到有机污染物分子上,这种反应被称为电化学氧转移反应(EOTR)。EOTR 的一个典型案例,是苯酚的阳极电解矿化:

$$C_6H_5OH + 11H_2O \rightarrow 6CO_2 + 28H^+ + 28e^- \tag{1.2}$$

在该反应中,参与苯酚加氧矿化过程的氧原子全部来源于水分子,而被释放的质子在阴极被还原发生析氢反应:

$$28H^+ + 28e^- \rightarrow 14H_2 \tag{1.3}$$

在析氧反应发生之前,理论上也有可能存在直接氧化过程,但反应速率通常较小,具体取决于阳极材料的电催化活性。Pt、Pd等贵金属以及IrO_2、Ru-TiO_2、Ir-TiO_2和PbO_2等金属氧化物具有较高的阳极反应活性。

然而,在析氧电位前的恒压电解过程中,电化学催化活性往往会逐渐降低(通常被称为中毒效应),这是由于在阳极表面形成了一层聚合物,导致电极失活。这种失活取决于三个因素:阳极表面的吸附特性、有机物与中间产物的浓度以及阳极附近的电解液环境。需要说明的是,在吸附性能较弱的阳极和惰性材料表面(如BDD、掺硼金刚石)[10],电催化活性的降低不太明显;在高浓度有机物或芳环类物质存在时,电极失活尤为显著,这类有机物主要包括:苯酚[11-14]、氯酚[15-17]、硝基取代酚[18-19]、苯胺[20]、萘酚[10]、芳香族和脂肪族烯烃[21]、除草剂[22]、氢醌[23]、合成染料[24]、吡啶[25]以及芳香磺酸盐等[26]。Rodrigo等人研究发现[27],在恒压条件下,水溶液中的4-氯酚(4-CP)可以在BDD阳极表面发生直接氧化。然而,直接电子转移反应会导致电极污染,因为在其表面形成了一层中间体聚合物膜。通过BDD电极在$5.0 \text{ mmol} \cdot L^{-1}$ 4-CP溶液中的循环伏安(CV)曲线可以看出,1.70 V(SHE)处的阳极氧化峰逐渐衰减并在第5圈彻底消失,由此证实了电极的失活(图1.2(a))。此处假设4-CP氧化过程中聚合物膜的形成是中间产物氯苯氧阳离子(或自由基)耦合的结果(图1.3)。

这种在BDD上形成聚合物的反应机理与Gattrell和Kirk提出的苯酚氧化过程中铂阳极失活的反应机理相似[13]。

Rodgers等人探讨了氯酚类物质电解过程中的阳极污染机理[17],并比较了不同氯化程度的苯酚的结构和反应活性。在$100 \text{ mV} \cdot s^{-1}$固定扫速下(图1.2(b)),4-氯酚在低浓度下的峰电流与$1.0 \text{ mmol} \cdot L^{-1}$以上浓度的峰电流不成线性比例,随着浓度的增加,峰电位逐渐降低。与此同时,水的氧化电位发生正移,这一结果与此前在大量电解苯酚过程中观察到的阳极污染现象是一致的[13,29],说明较低的浓度会使发生电极污染的电位升高(难度增大)。此外,他们观察到,在线性扫描伏安(LSV)谱图中,酚的氯代程度越高,阳极峰电位值反而会随着氯原子数量的增加而变小。

中毒效应可以通过析氧或间接氧化来去除有机聚合物以达到"解毒",但这种"解毒"过程在低偏压(即发生析氧反应前的低外加偏压)条件下难以实施,因此成为直接氧化的技术瓶颈。

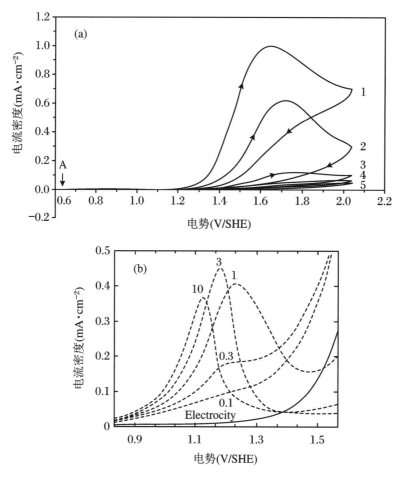

图1.2 (a) BDD 在 5.0 mmol·L^{-1} 4-氯酚(1.0 mol·L^{-1} H$_2$SO$_4$)中循环5次的 CV 扫描曲线;(b) Pt 在不同浓度 4-氯酚(0.1~10 mmol·L^{-1},pH=6.0)中的 LSV 扫描曲线[28]

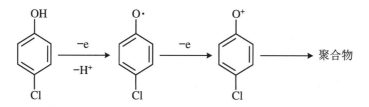

图1.3 4-氯酚氧化过程中聚合物膜的形成[28]

1.1.1.2 间接氧化

为了防止电极污染,间接氧化是一个可行的思路,能够有效避免有机物与阳极表面的直接电子交换。在间接氧化过程中,污染物与通过一些电化学反应产生的氧化还原中间体活性物质而发生氧化,这些活性物质充当着污染物与阳极发生电子交换的媒介。在间接氧化过程中获得高效率的主要要求如下[3]:

(1) 活性物质产生的电位不能接近析氧电位;
(2) 活性物质生成率高;
(3) 活性物质和污染物的反应速率高于任何竞争性反应的速率;
(4) 尽量减少污染物的吸附。

在间接氧化体系中,无需向溶液中添加氧化催化剂,也不产生任何副产物。但二次析氧反应的发生会降低电流效率,这成为间接氧化的主要技术问题。

Johnson 等人推测[30-40],在高阳极电位下的氧转移反应涉及从水分解所产生的吸附态·OH 自由基:

$$S[\ ] + H_2O \rightarrow S[\cdot OH] + H^+ + e^- \tag{1.4}$$

$$S[\cdot OH] + R \rightarrow S[\ ] + RO + H^+ + e^- \tag{1.5}$$

式中,S 为吸附态·OH 的表面位置。而这个反应不可避免地存在析氧副反应:

$$S[\cdot OH] + H_2O \rightarrow S[\ ] + O_2 + 3H^+ + 3e^- \tag{1.6}$$

Comninellis 等人发现[41-44],电极材料的性质对反应的选择性和效率都有很大的影响。为了解释这些现象,他们提出了一种同时进行析氧与有机物氧化的金属氧化物综合模型,如图 1.4 所示。

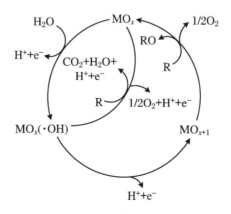

图 1.4 "活性"与"非活性"阳极材料对污染物的阳极氧化[28]

与Johnson提出的机理类似,氧转移反应的第一步是分解水形成吸附态的·OH自由基:

$$MO_x + H_2O \rightarrow MO_x(\cdot OH) + H^+ + e^- \tag{1.7}$$

以上步骤取决于电极材料的性质,根据电极材料的性质有可能区分为两种极端条件下的电极,即"活性"和"非活性"阳极。

在"活性"阳极上,当电极表面有较高的氧化态时,被吸附的·OH自由基可能与阳极相互作用,形成所谓的高价氧化物:

$$MO_x(\cdot OH) \rightarrow MO_{x+1} + H^+ + e^- \tag{1.8}$$

表面氧化还原电对MO_{x+1}/MO_x,有时被称为化学吸附的"活性氧",可作为"活性"电极上有机物转化或选择性氧化的媒介:

$$MO_{x+1} + R \rightarrow MO_x + RO \tag{1.9}$$

在"非活性"阳极上,排除了高价氧化物的形成,·OH——物理吸附的"活性氧",大概率会参与有机物的非选择性氧化,使之完全矿化为CO_2:

$$MO_x(\cdot OH) + R \rightarrow MO_x + H_2O + CO_2 + H^+ + e^- \tag{1.10}$$

然而,无论是化学吸附还是物理吸附的"活性氧"都会发生竞争性的副反应,即析氧反应,导致阳极电流效率降低。

如表1.1所示,对于碳、石墨、IrO_2、RuO_2和Pt等低析氧过电位的阳极,电极具有"活性"行为,容易发生析氧反应,电压稍高则析氧反应显著增加,因而只能使有机物发生部分氧化;而例如Sb-SnO_2、PbO_2、BDD等高析氧过电位的阳极,它们具有"非活性"行为,有利于污染物的完全矿化,因此成为废水处理的理想电极。然而在实际反应过程中,大多数阳极会表现出混合性能,即两个平行的反应路径都发生了污染物氧化和析氧反应。

表1.1 各阳极材料在H_2SO_4中的析氧电位(标准电位为1.23 V/SHE)[28]

阳极材料	析氧电位(V/SHE)	电解质浓度
RuO_2	1.47	0.50 mol·L^{-1} H_2SO_4
IrO_2	1.52	0.50 mol·L^{-1} H_2SO_4
Pt	1.60	0.50 mol·L^{-1} H_2SO_4
定向热解石墨	1.70	0.50 mol·L^{-1} H_2SO_4
SnO_2	1.80	0.05 mol·L^{-1} H_2SO_4
PbO_2	1.90	1.00 mol·L^{-1} H_2SO_4
BDD	2.30	0.50 mol·L^{-1} H_2SO_4

1.1.2
电化学技术所面临的问题和挑战

任何技术方法都不可能十全十美。正如之前所提到的,电化学水处理技术在具备显著优势的同时,也存在着明显的缺陷和问题。

1. 电催化机理的不确定性

阳极氧化是一个非常复杂的反应过程,电极材料、反应条件和溶液组分的多样性,导致该过程往往为直接氧化和间接氧化共存的平行反应路径[28],这对解析电催化过程的反应机理造成了极大的困难。

2. 技术瓶颈

直接氧化容易发生阳极污染,间接氧化电流效率较低[28]。

3. 优异电极材料缺失

$Sb-SnO_2$、PbO_2、BDD 虽然可以作为优异的废水处理材料,但 $Sb-SnO_2$ 的污染物矿化能力有限,PbO_2 具有较高的毒性,而 BDD 的价格又过于昂贵,所以它们均不适合作为电化学水处理材料进行推广和应用。

因此,发展实用、前景广阔、具有明确电催化机理的污染控制材料成为环境领域科研人员的重要研究目标。

1.2
TiO_2 在环境领域中的研究现状

1.2.1
材料学优势及在环境领域中的应用

TiO_2 是一种储量丰富、成本低廉、化学性质稳定且环境友好的绿色无机材料。它已被成功商业化，目前广泛应用于涂料、造纸、印刷油墨、橡胶和化妆品等的生产中[45]，也成为各学科领域的研究热点。在 TiO_2 的诸多应用研究中，光催化最具有代表性。

半导体光催化技术能够有效降解有机污染物，其体系成本低、环境友好且操作条件温和，因而成为去除有机污染物的方法之一[46-49]。光催化技术以太阳能为驱动力，利用半导体纳米材料将光能转化为化学能，来降解矿化有机物，适用于水处理和废气处理[50]。

半导体光催化剂的催化机理是基于能带理论建立的。典型的反应过程可分为三个阶段[51-52]。首先，半导体材料吸收入射光，价带（valence band）中的电子（e^-）获得能量后受激发跃迁至导带（conduction band），同时在价带产生一个带正电的空穴（h^+）；然后，催化剂中部分导带的光生电子会与价带的光生空穴发生复合并以热辐射的形式将吸收的光能重新释放出来，而另一部分光生电子和空穴则分别沿着晶格进行输运，直至到达催化剂表面，这一过程称为载流子（即空穴与电子）的分离；最后，部分到达催化剂表面的空穴和电子注入吸附在催化剂表面的分子中，从而分别引发氧化（h^+）和还原（e^-）反应。

1972年，两位日本科学家 Fujishima 和 Honda 发现了紫外光照射下 TiO_2 晶体表面发生的水分解现象，"光催化"这一概念从此进入人们的视野[53]。之后的几十年里，TiO_2 成为了研究最广泛、最深入的光催化剂材料，应用领域覆盖化学燃料生产、空气和水的净化、表面的自净和消毒等[50]。

由于 TiO_2 的禁带宽度普遍大于 3.0 eV，因此常规的 TiO_2 催化剂通常只能在紫外光照射下产生催化活性[54]。由于太阳光中的紫外光含量很少（约5%），TiO_2 在太阳光照射下的催化效率往往很低[54-55]。为了解决这个问题，人们进行了大量研究工作以期提升 TiO_2 的可见光（vis）催化效率[56]。其中，氮、碳等阴离子掺杂能够有效降低 TiO_2 的禁带宽度，从而提升其可见光响应[57-58]。此外，调控晶型比例形成同质结并结合硼掺杂进行修饰，也可以提升 TiO_2 的光催化效率[59]。

1.2.2

光催化技术缺陷

虽然光催化技术在保护环境方面具有一定的应用前景，但也存在严重的技术缺陷：

(1) 光转换效率低,每一步光转换过程都伴随着能量效率的损失[60-61]。

(2) 天然有机质(NOM)、碳酸盐和其他背景成分的存在,容易降低光吸收率、淬灭活性氧自由基,从而使光催化过程的有效性严重受限[62-64]。

综上所述,光催化技术在环境领域中的实际应用往往受到很大程度的限制,如图1.5所示[65]。因此,TiO_2在环境水处理领域的新应用亟待研发,而其在电化学技术中的应用与改良成为了一个可能的突破口。

1.3 TiO_2电化学水处理的应用挑战与解决策略

虽然TiO_2在光催化体系中具有良好的催化活性[44,66],但是由于其电导率低、阳极活性差[67-68],通常不能作为有效的电催化剂[68-69],因此也难以在电化学水处理中发挥效力。如何提升TiO_2的电导率和阳极活性,是拓展其电化学水处理功能所需解决的重要问题。

通常来说,良好的电催化电极材料应具备如下性能:

(1) 良好的导电性。至少与导电基底材料(如碳纸)结合后能为电子交换反应提供不引起严重电压降的电子通路,即电极材料的电阻不能太大。

(2) 高的催化活性。能够实现目标催化反应,抑制不需要或有害的副反应。

(3) 良好的稳定性。能够耐受杂质和中间产物的作用而不致于较快地被污染(或中毒)而失活,并且在实现催化反应的电压范围内催化表面不会因电化学反应而过早失去催化活性。

(4) 良好的机械物理性质。表面不脱层、不溶解。

在上述性能中,电极材料的催化活性至关重要。目前,通常有两种策略来提升电催化体系的活性(或反应速率):① 增加指定电极上催化活性位点的数量;② 增加每个催化活性位点的本征活性。这些策略(图1.6)可以在合理调控下同时实施,从而最大限度地改善电极的催化性能[70]。此外,在不影响其他重要过程(如电荷和质量传输)的情况下,电极的催化剂负载量存在物理限制,因此只有

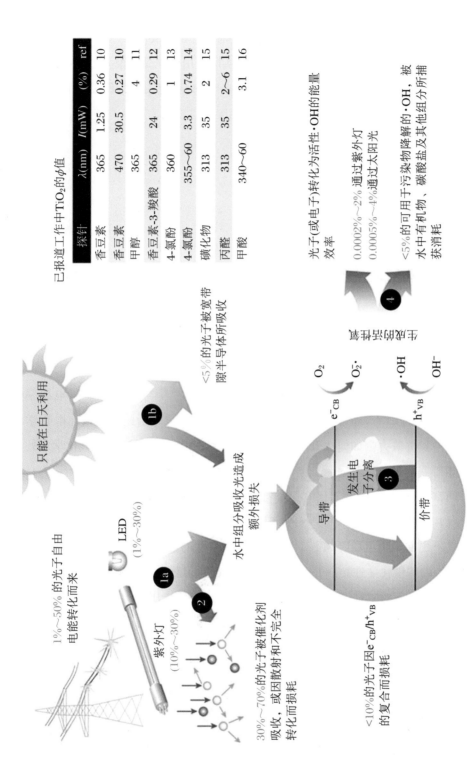

图1.5 光催化技术的主要缺陷[65]

改善材料的本征活性才能充分提升电极活性,从而减轻高催化剂负载量引起的传质问题。

综合上述考虑,可以采用如下具体策略,来尝试解决 TiO_2 电导率低、阳极活性差的问题。

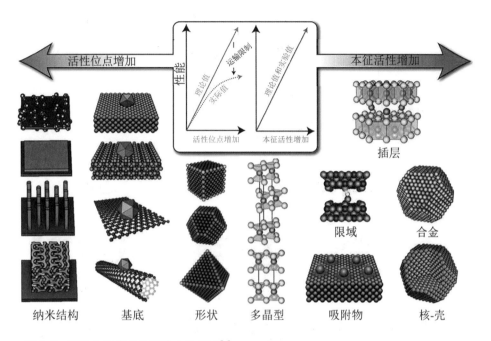

图 1.6　提升电极催化性能的方法策略[2]

1.3.1
单晶设计与晶面调控

在不改变原子种类的条件下,形貌、晶体结构和暴露晶面等参数会对 TiO_2 的物理化学性质及催化性能产生很大影响。因单晶材料内部晶体结构连续有序,具有较高的导电性,因此合成 TiO_2 单晶成为提升其电导率的有效方法之一。

除了设计单晶结构之外,特异晶相的选择、高活性晶面的暴露也可能是提升催化剂本征活性的良好手段。TiO_2 具有三种主要的晶相结构:锐钛矿、金红石和板钛矿,前两者最为常见。由于锐钛矿具有较高的催化反应活性,因而被研究得最多。锐钛矿 TiO_2 晶相存在三种常见的晶面:{001}、{100}、{101}(图 1.7),它们的平均表面能分别为:{001} = 0.90 J·m^{-2} > {100} = 0.53 J·m^{-2} > {101}

$=0.44 \text{ J} \cdot \text{m}^{-2}$[15]。在 TiO_2 的{001}晶面上,表面原子的排列和协调具有独特的原子结构和电子结构,从而决定了其良好的化学稳定性、吸附性能和催化反应活性。因此,实现锐钛矿{001}晶面的高暴露,是提升 TiO_2 电催化活性的重要方法。

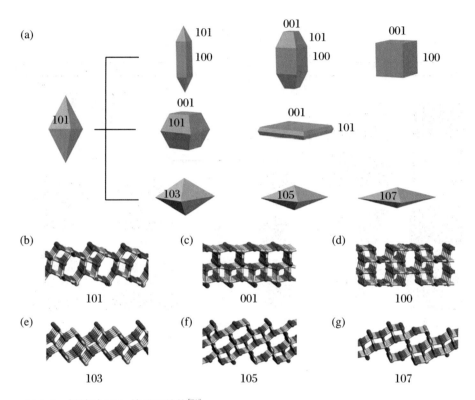

图 1.7　锐钛矿 TiO_2 的晶面结构[71]

然而在实际合成中,更容易在锐钛矿 TiO_2 表面暴露的是热力学性能稳定的低能{101}晶面(根据 Wulff 结构,超过 94%),因而催化活性较低[72]。为了提高其反应性能,可以采取晶面调控措施,以制备出高暴露{001}晶面的锐钛矿 TiO_2。

切角八面体双锥(TOB,图 1.8(左)),有 8 个{101}晶面和 2 个{001}晶面(上下顶角切面),是在自然界中观察到的一种非常普遍的晶体形状[74],也是基于 Wulff 结构的最常见的锐钛矿晶形[75]。在正常情况下,截断度(B/A)为 0.3～0.4,所暴露的{001}晶面不到 10%[75]。

为了合成高能{001}晶面暴露的锐钛矿 TiO_2 晶体(图 1.8(右)),在非平衡条件下,晶体生长应限制在动力学控制范围内[76-77]。例如,$TiCl_4$ 或 $Ti(OC_4H_9)_4$ 的高温(1200 ℃)气相氧化,通过快速的加热和淬火产生非平衡条件,可以合成近 40%{001}暴露晶面的锐钛矿 $TOBs$[67-78]。

大多数钛系材料都是在水或非水的液相环境中合成的。通过选择合适的钛前驱体、反应介质(溶剂)和封盖剂,并结合控制反应温度和压力,可以很容易地控制反应动力学和成核过程。影响单晶生长模式的一个关键参数就是晶体的表面能[77]。因为晶体生长速率与晶体表面能呈指数关系,所以通过对比表面进行改性来提高或降低表面能,可以调控晶体生长和目标晶体形状。

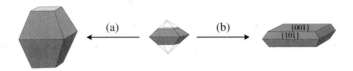

图 1.8 锐钛矿 TiO_2 的晶体演化[73]:(a) 在平衡条件下,高能{001}晶面迅速减少,晶体自发转变成具有热力学稳定性的以{101}为主的 TOB(左);(b) 在非平衡条件下,高能{001}切面可以(通过形貌控制剂和晶体生长)被稳定下来,形成以亚稳态{001}为主的 TOB(右)

一种通过改变表面能来控制最终形状的普遍方法,是在有吸附物质存在的情况下生长晶体,这些吸附物质与不同的晶面有不同的相互作用[75],从而在晶体表面发生动态溶剂化过程和特定封盖剂的选择性黏附作用[76]。不同切面的比率决定了切面的相对稳定性,因而吸附剂的性质和浓度能够确定最终的晶体形状。一个典型的例子是氟与锐钛矿 TiO_2 晶体{001}晶面的相互作用。在理论计算结果的辅助支撑下,Yu 等人首先指出,吸附的氟原子可以使末端为氟的锐钛矿 TiO_2 晶体中的{001}晶面异常稳定(图 1.9(a、b))[80]。随后,他们在特定的水热条件下,以 HF 为封盖剂、TiF_4 为前驱体,成功合成了{001}晶面暴露率为 47%的锐钛矿 TOBs 单晶(图 1.9(c)),证实了此前的预测。这一开创性工作引发了学术界对高能{001}晶面暴露锐钛矿 TiO_2 晶体合成的广泛关注。经过深入研究发现,{001}暴露晶面比例的增加,得益于材料粒径的减小和比表面积的增大[72-73,80-87]。在其他钛源(如 TiN[72]、TiB_2[73] 和 TiC[80] 等)的实验尝试中,只有在少数情况下能够成功合成具有纳米尺寸的 TOBs(表 1.2)[81-83]。Xie 等人将钛酸四丁酯与高浓度 HF 溶液混合后进行水热反应,合成了具有 89%的{001}晶面暴露率的锐钛矿 TiO_2 纳米薄片(图 1.9(d))[82]。

需要说明的是,HF 是最常用的封盖剂,但因为 HF 液体和蒸气都具有强腐蚀性和高毒性,不适合大规模生产,因此溶剂热法应运而生,通常采用低浓度氢氟酸与乙醇或离子液体(ILs)结合使用(表 1.2)[84-86]。Lu 等人证明,在该方法中,{001}晶面暴露率可以通过乙醇和氟的协同作用在一定程度上继续增加[84],而乙醇的使用还能够更加灵活地控制分散程度和颗粒大小[85]。此外,醇类可以

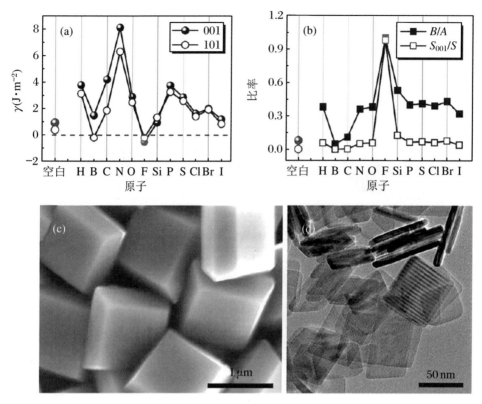

图 1.9 氟介导的锐钛矿{001}晶面的形貌参数[73]

在一定程度上调控一维纳米结构的生长,从而能够合成指定形貌的多维超结构[86]。

表 1.2 几种典型{001}晶面暴露锐钛矿 TiO_2 薄片的水热/溶剂热合成条件[73]

前驱体	封盖剂	溶剂	尺寸×厚度	{001}暴露率
TiF_4/TiN/TiB_2/TiS_2/TiC/Ti-powder	HF/NH_4HF_2/NH_4F	水	(1~5) μm×1 μm	40%~60%
$Ti(SO_4)_2$/TBOT	HF	水	50 nm×10 nm	18%~89%
TiF_4/$TiCl_4$	HF	水-醇	1.09 mm×260 nm	50%~65%
TiF_4/$TiCl_4$	HF	水	微米级	27%~80%

上述合成过程主要在水溶剂中进行。虽然在非水溶剂体系合成锐钛矿 TiO_2 SCs 的研究已经取得了良好的进展,但是选择合适的封盖剂对晶体形状的设计仍然具有挑战性[63,88]。

1.3.2

缺陷工程

缺陷工程是一种通过调节中心金属晶格的原子和电子结构以达到高活性的结构自修饰方法[89],能够使过渡金属氧化物的电催化活性得到显著提高。

大量研究表明,具有局域无序原子和不规则结构的氧空位可以作为催化活性位点,通过增加费米能级附近的态密度进行有效的电子转移、弱化晶格的金属-氧键,实现热力学多相催化的快速氧原子交换,使过渡金属氧化物的电催化活性得到显著提高[89-93]。其中,Magneli 相 TiO_2 伴随着有序的晶体重排和极小的点缺陷(Ti_nO_{2n-1}),被证明是一种经济且有应用前景的电极材料,能够用于信息存储、能量存储转换和水处理[91]。此外,在能源和环境应用中,非化学计量的金属氧化物通常也能在 O_2/CO_2 还原、催化产氢和污染物降解方面表现出优于其本征氧化物的性能[93-96]。

因此,缺陷工程也可以作为提升 TiO_2 电催化活性的辅助方法之一。

本章小结

如上所述,电化学技术是一种有效的水处理方法,而 TiO_2 具备开发成为电催化材料的良好潜质,针对其面临的主要问题和挑战,本章提出了可能的解决方案。本书将针对这些具体方案进行逐一讨论,着重分析 TiO_2 的电催化氧化、还原反应机理,同时对催化体系进行改良优化,力争实现污染物的高效降解去除。

参考文献

[1] Bousher A, Shen X, Edyvean R. Removal of coloured organic matter by adsorption onto low-cost waste materials[J]. Water Res., 1997, 31(8): 2084-2092.

[2] Seh Z W, Kibsgaard J, Dickens C F, et al. Combining theory and experiment in electrocatalysis: Insights into materials design[J]. Science, 2017, 355(6321): eaad4998.

[3] Rajeshwar K, Ibanez J G, Swain G M. Electrochemistry and the environment[J]. J. Appl. Electrochem., 1994, 24(12): 1077-1091.

[4] Grimm J, Bessarabov D, Sanderson R. Review of electro-assisted methods for water purification[J]. Desalination, 1998, 115: 285-294.

[5] Janssen L J J, Koene L. The role of electrochemistry and electrochemical technology in environmental protection[J]. Chem. Eng. J., 2002, 85(2/3): 137-146.

[6] Juettner K, Galla U, Schmieder H. Electrochemical approaches to environmental problems in the process industry[J]. Electrochim. Acta, 2000, 45: 2575-2594.

[7] Simonsson D. Electrochemistry for a cleaner environment[J]. Chem. Soc. Rev., 1997, 26: 181-189.

[8] Martínez-Huitle C A, Ferro S. Electrochemical oxidation of organic pollutants for the wastewater treatment: direct and indirect processes[J]. Chem. Soc. Rev., 2006, 12: 1324-1340.

[9] Chen G. Electrochemical technologies in wastewater treatment[J]. Sep. Purif. Technol., 2004, 38: 11-41.

[10] Panizza M, Cerisola G. Influence of anode material on the electrochemical oxidation of 2-naphthol. Part 1: Cyclic voltammetry and potential step experiments[J]. Electrochim. Acta, 2003, 48: 3491-3497.

[11] Canizares P, Martinez F, Diaz M, et al. Electrochemical oxidation of aqueous phenol wastes using active and nonactive electrodes[J]. J. Electrochem. Soc., 2002, 149: D118-D124.

[12] Foti G, Gandini D, Comninellis C. Anodic oxidation of organics on thermally prepared oxide electrodes[J]. Curr. Top. Electrochem., 1997, 5: 71-91.

[13] Gattrell M, Kirk D. A Study of the Oxidation of Phenol at Platinum and Preoxidized Platinum Surfaces[J]. J. Electrochem. Soc., 1993, 140: 1534-1540.

[14] Iniesta J, Michaud P A, Panizza M, et al. Electrochemical oxidation of phenol at boron-doped diamond electrode[J]. Electrochim. Acta, 2001, 46:

3573-3578.

[15] Rodrigo M A, Michaud P A, Duo I, et al. Oxidation of 4-chlorophenol at boron-doped diamond electrode for wastewater treatment [J]. J. Electrochem. Soc., 2001, 148: D60-D64.

[16] Rao T N, Terashima C, Sarada B V, et al. Electrochemical oxidation of chlorophenols at a boron-doped diamond electrode and their determination by high-performance liquid chromatography with amperometric detection [J]. Anal. Chem., 2002, 74: 895-902.

[17] Rodgers J D, Jedral W, Bunce N J. Electrochemical oxidation of chlorinated phenols[J]. Environ. Sci. Technol., 1999, 33: 1453-1457.

[18] Canizares P, Saez C, Lobato J, et al. Electrochemical treatment of 4-nitrophenol-containing aqueous wastes using boron-doped diamond anodes [J]. Ind. Eng. Chem. Res., 2004, 43, 1944-1951.

[19] Canizares P, Saez C, Lobato J, et al. Electrochemical treatment of 2,4-dinitrophenol aqueous wastes using boron-doped diamond anodes [J]. Electrochim. Acta, 2004, 49: 4641-4650.

[20] Mitadera M, Spataru N, Fujishima A. Electrochemical oxidation of aniline at boron-doped diamond electrodes[J]. J. Appl. Electrochem., 2004, 34: 249-264.

[21] Zanta C L P S, De Andrade A R, Boodts J F C. Electrochemical behaviour of olefins: oxidation at ruthenium-titanium dioxide and iridium-titanium dioxide coated electrodes[J]. J. Appl. Electrochem., 2000, 30: 467-474.

[22] Boye B, Brillas E, Marselli B, et al. Electrochemical incineration of chloromethylphenoxy herbicides in acid medium by anodic oxidation with boron-doped diamond electrode [J]. Electrochim. Acta, 2006, 51: 2872-2880.

[23] Nasr B, Abdellatif G, Canizares P, et al. Electrochemical oxidation of hydroquinone, resorcinol, and catechol on boron-doped diamond anodes[J]. Environ. Sci. Technol., 2005, 39: 7234-7239.

[24] Saez C, Panizza M, Rodrigo M A, et al. Electrochemical incineration of dyes using a boron-doped diamond anode [J]. J. Chem. Technol. Biotechnol., 2007, 82: 575-581.

[25] Iniesta J, Michaud P A, Panizza M, et al. Electrochemical oxidation of 3-methylpyridine at a boron-doped diamond electrode: application to

electroorganic synthesis and wastewater treatment[J]. Electrochem. Commun., 2001, 3: 346-351.

[26] Panizza M, Cerisola G. Electrochemical oxidation of aromatic sulfonated acids on a boron-doped diamond electrode[J]. Int. J. Environ. Pollut., 2006, 27: 64-74.

[27] Katz A, McDonagh A, Tijing L, et al. Fouling and inactivation of titanium dioxide-based photocatalytic systems[J]. Environ. Sci. Technol., 2015, 45: 1880-1915.

[28] Panizza M, Cerisola G. Direct and mediated anodic oxidation of organic pollutants[J]. Chem. Rev., 2009, 109: 6541-6569.

[29] Comninellis C, Vercesi G P. Characterization of DSA$^©$-type oxygen evolving electrodes: choice of a coating[J]. J. Appl. Electrochem., 1991, 21: 335-345.

[30] Chang H, Johnson D C. Electrocatalysis of Anodic Oxygen-Transfer Reactions Activation of PbO_2-Film Electrodes in 1.0 mol·L^{-1} $HClO_4$ by Addition of Bismuth(Ⅲ) and Arsenic(Ⅲ, Ⅴ)[J]. J. Electrochem. Soc., 1990, 137: 2452-2457.

[31] Chang H, Johnson D C. Electrocatalysis of anodic oxygen-transfer reactions ultrathin films of lead oxide on solid electrodes[J]. J. Electrochem. Soc., 1990, 137: 3108-3113.

[32] Feng J, Johnson D C. Electrocatalysis of anodic oxygen-transfer reactions fe-doped beta-lead dioxide electrodeposited on noble metals[J]. J. Electrochem. Soc., 1990, 137: 507-510.

[33] Feng J, Johnson D C. Electrocatalysis of anodic oxygen-transfer reactions: titanium substrates for pure and doped lead dioxide films[J]. J. Electrochem. Soc., 1991, 138: 3328-3337.

[34] Feng J, Houk L L, Johnson D C, et al. Electrocatalysis of anodic oxygen-transfer reactions: the electrochemical incineration of benzoquinone[J]. J. Electrochem. Soc., 1995, 142: 3626-3632.

[35] Houk L L, Johnson S K, Feng J, et al. Electrochemical incineration of benzoquinone in aqueous media using a quaternary metal oxide electrode in the absence of a soluble supporting electrolyte[J]. J. Appl. Electrochem., 1998, 28: 1167-1177.

[36] Johnson S K, Houk L L, Feng J, et al. Electrochemical incineration of

4-chlorophenol and the identification of products and intermediates by mass spectrometry[J]. Environ. Sci. Technol., 1999, 33: 2638-2644.

[37] Kawagoe K T, Johnson D C. Electrocatalysis of anodic oxygen-transfer reactions oxidation of phenol and benzene at bismuth-doped lead dioxide electrodes in acidic solutions[J]. J. Electrochem. Soc., 1994, 141: 3404.

[38] Treimer S E, Feng J, Scholten M C, et al. Cmparison of voltammetric responses of toluene and xylenes at iron(Ⅲ)-doped, bismuth(Ⅴ)-doped, and undoped β-lead dioxide film electrodes in 0.50 mol·L^{-1} H_2SO_4[J]. J. Electrochem. Soc., 2001, 148: E459-E463.

[39] Vitt J E, Johnson D C. The importance of anodic discharge of H_2O in anodic oxygen-transfer reactions[J]. J. Electrochem. Soc., 1992, 139: 774-778.

[40] Feng J, Johnson D C, Lowery S N, et al. Electrocatalysis of anodic oxygen-transfer reactions evolution of ozone[J]. J. Electrochem. Soc., 1994, 141: 2708-2711.

[41] Simond O, Schaller V, Comninellis C. Theoretical model for the anodic oxidation of organics on metal oxide electrodes[J]. Electrochim. Acta, 1997, 42: 2009-2012.

[42] Foti G, Gandini D, Comninellis C, et al. Oxidation of organics by intermediates of water discharge on IrO_2 and synthetic diamond anodes[J]. Electrochem. Solid State, 1999, 2(5): 228-230.

[43] Comninellis C. Electrocatalysis in the electrochemical conversion/combustion of organic pollutants for waste water treatment[J]. Electrochim. Acta, 1994, 39: 1857-1862.

[44] Comninellis C, De Battisti A. Electrocatalysis in anodic oxidation of organics with simultaneous oxygen evolution[J]. J. Chim. Phys., 1996, 93: 673-679.

[45] Chen X B, Mao S S. Titanium dioxide nanomaterials: synthesis, properties, modifications, and applications[J]. Chem. Rev., 2007, 107: 2891-2959.

[46] Linsebigler A L, Lu G, Yates Jr J T. Photocatalysis on TiO_2 surfaces: principles, mechanisms, and selected results[J]. Chem. Rev., 1995, 95: 735-758.

[47] Dalrymple O K, Stefanakos E, Trotz M A, et al. A review of the mechanisms and modeling of photocatalytic disinfection[J]. Appl. Catal.,

B, 2010, 98: 27-38.

[48] Thompson T L, Yates J T. Surface science studies of the photoactivation of TiO$_2$ new photochemical processes[J]. Chem. Rev., 2006, 106: 4428-4453.

[49] Serpone N, Emeline A. Semiconductor photocatalysis: past, present, and future outlook[J]. J. Phys. Chem. Lett., 2012, 3(5): 673-677.

[50] Fujishima A, Zhang X, Tryk D A. TiO$_2$ photocatalysis and related surface phenomena[J]. Surf. Sci. Rep., 2008, 63: 515-582.

[51] Kubacka A, Fernandez-Garcia M, Colon G. Advanced nanoarchitectures for solar photocatalytic applications [J]. Chem. Rev., 2011, 112: 1555-1614.

[52] Chen C, Ma W, Zhao J. Semiconductor-mediated photodegradation of pollutants under visible-light irradiation[J]. Chem. Soc. Rev., 2010, 39: 4206-4219.

[53] Fujishima A, Honda K. Electrochemical photolysis of water at a semiconductor electrode[J]. Nature, 1972, 238: 37-38.

[54] Tao J, Luttrell T, Batzill M. A two-dimensional phase of TiO$_2$ with a reduced bandgap[J]. Nat. Chem., 2011, 3: 296-300.

[55] Bak T, Nowotny J, Rekas M, et al. Photo-electrochemical hydrogen generation from water using solar energy. Materials-related aspects[J]. Int. J. Hydrogen Energy, 2002, 27: 991-1022.

[56] Choi W, Termin A, Hoffmann M R. The role of metal ion dopants in quantum-sized TiO$_2$: correlation between photoreactivity and charge carrier recombination dynamics[J]. J. Phys. Chem., 1994, 98: 13669-13679.

[57] Asahi R, Morikawa T, Ohwaki T, et al. Visible-light photocatalysis in nitrogen-doped titanium oxides[J]. Science, 2001, 293: 269-271.

[58] Khan S U, Al-Shahry M, Ingler W B. Efficient photochemical water splitting by a chemically modified n-TiO$_2$ [J]. Science, 2002, 297: 2243-2245.

[59] Wang W K, Chen J J, Gao M, et al. Photocatalytic degradation of atrazine by boron-doped TiO$_2$ with a tunable rutile/anatase ratio[J]. Appl. Catal., B, 2016, 195: 69-76.

[60] Loeb S K, Alvarez P J J, Brame J A, et al. The technology horizon for photocatalytic water treatment: sunrise or sunset? [J]. Environ. Sci. Technol., 2019, 53: 2937-2947.

[61] Mierzwa J C, Rodrigues R, Teixeira A C S C. UV-hydrogen peroxide processes[J]. Adv. Oxid. Processes Wastewater Treat., 2018: 13-48.

[62] Liao C H, Gurol M D. Chemical oxidation by photolytic decomposition of hydrogen peroxide[J]. Environ. Sci. Technol., 1995, 29: 3007-3014.

[63] Haag W R, Yao C C D. Rate constants for reaction of hydroxyl radicals with several drinking water contaminants[J]. Environ. Sci. Technol., 1992, 26: 1005-1013.

[64] Brame J, Long M, Li Q, et al. Inhibitory effect of natural organic matter or other background constituents on photocatalytic advanced oxidation processes: mechanistic model development and validation[J]. Water Res., 2015, 84: 362-371.

[65] Benotti M J, Stanford B D, Wert E C, et al. Evaluation of a photocatalytic reactor membrane pilot system for the removal of pharmaceuticals and endocrine disrupting compounds from water[J]. Water Res., 2009, 43: 1513-1522.

[66] Liu G, Sun C H, Smith S C, et al. Sulfur doped anatase TiO_2 single crystals with a high percentage of {001} facets[J]. J. Colloid Interface Sci., 2010, 349: 477-483.

[67] Ahonen P P, Moisala A, Tapper U, et al. Gas-phase crystallization of titanium dioxide nanoparticles[J]. J. Nanopart. Res., 2002, 4: 43-52.

[68] Wu B H, Guo C Y, Zheng N F, et al. Nonaqueous production of nanostructured anatase with high-energy facets[J]. J. Am. Chem. Soc., 2008, 130: 17563-17567.

[69] Dinh C T, Nguyen T D, Kleitz F, et al. Shape-controlled synthesis of highly crystalline titania nanocrystals[J]. ACS Nano, 2009, 3: 3737-3743.

[70] Benck J D, Hellstern T R, Kibsgaard J, et al. Catalyzing the Hydrogen Evolution Reaction (HER) with molybdenum sulfide nanomaterials[J]. ACS Catal., 2004, 4: 3957-3971.

[71] Wen Z Z, Jiang H B, Qiao S Z, et al. Synthesis of high-reactive facets dominated anatase TiO_2[J]. J. Mater. Chem., 2011, 21: 7052-7061.

[72] Lazzeri M, Vittadini A, Selloni A. Structure and energetics of stoichiometric TiO_2 anatase surfaces[J]. Phys. Rev. B: Condens. Matter, 2001, 63: 155409.

[73] Liu G, Yang H G, Wang X W, et al. Enhanced photoactivity of oxygen-

deficient anatase TiO$_2$ sheets with dominant {001} facets[J]. J. Phys. Chem. C., 2009, 113: 21784-21788.

[74] Diebold U. The surface science of titanium dioxide[J]. Surf. Sci. Rep., 2003, 48: 53-229.

[75] Selloni A. Anatase shows its reactive side[J]. Nat. Mater., 2008, 7: 613-615.

[76] Yin Y, Alivisatos A P. Colloidal nanocrystal synthesis and the organic-inorganic interface[J]. Nature, 2005, 437: 664-670.

[77] Jun Y W, Lee J H, Choi J S, et al. Symmetry-controlled colloidal nanocrystals: nonhydrolytic chemical synthesis and shape determining parameters[J]. J. Phys. Chem. B, 2005, 109: 14795-14806.

[78] Amano F, Prieto-Mahaney O O, Terada Y, et al. Decahedral single-crystalline particles of anatase titanium (IV) oxide with high photocatalytic activity[J]. Chem. Mater., 2009, 21: 2601-2603.

[79] Yang H G, Sun C H, Qiao S Z, et al. Anatase TiO$_2$ single crystals with a large percentage of reactive facets[J]. Nature, 2008, 453: 638-641.

[80] Yu J G, Dai G P, Xiang Q J, et al. Fabrication and enhanced visible-light photocatalytic activity of carbon self-doped TiO$_2$ sheets with exposed {001} facets[J]. J. Mater. Chem., 2011, 21: 1049-1057.

[81] Liu G, Sun C H, Yang H G, et al. Nanosized anatase TiO$_2$ single crystals for enhanced photocatalytic activity[J]. Chem. Commun., 2010, 46: 755-757.

[82] Han X G, Kuang Q, Jin M S, et al. Synthesis of titania nanosheets with a high percentage of exposed {001} facets and related photocatalytic properties[J]. J. Am. Chem. Soc., 2009, 131: 3152-3153.

[83] Dai Y Q, Cobley C M, Zeng J, et al. Synthesis of anatase TiO$_2$ nanocrystals with exposed {001} facets[J]. Nano Lett., 2009, 9: 2455-2459.

[84] Yang H G, Liu G, Qiao S Z, et al. Solvothermal synthesis and photoreactivity of anatase TiO$_2$ nanosheets with dominant {001} facets[J]. J. Am. Chem. Soc., 2009, 131: 4078-4083.

[85] Zhu J, Wang S H, Bian Z F, et al. Solvothermally controllable synthesis of anatase TiO$_2$ nanocrystals with dominant {001} facets and enhanced photocatalytic activity[J]. CrystEngComm, 2010, 12: 2219-2224.

[86] Zheng Z K, Huang B B, Qin X Y, et al. Highly efficient photocatalyst:

TiO₂ microspheres produced from TiO₂ nanosheets with a high percentage of reactive {001} facets[J]. Chem. Eur. J., 2009, 15: 12576-12579.

[87] Chen J S, Tan Y L, Li C M, et al. Constructing hierarchical spheres from large ultrathin anatase TiO₂ nanosheets with nearly 100% exposed {001} facets for fast reversible lithium storage[J]. J. Am. Chem. Soc., 2010, 132: 6124-6130.

[88] Jun Y W, Casula M F, Sim J H, et al. Surfactant-assisted elimination of a high energy facet as a means of controlling the shapes of TiO₂ nanocrystals [J]. J. Am. Chem. Soc., 2003, 125: 15981-15985.

[89] Chen D J, Chen C, Baiyee Z M, et al. Nonstoichiometric oxides as low-cost and highly-efficient oxygen reduction/evolution catalysts for low-temperature electrochemical devices[J]. Chem. Rev., 2015, 115: 9869-9921.

[90] Jing Y, Almassi S, Mehraeen S, et al. The roles of oxygen vacancies, electrolyte composition, lattice structure, and doping density on the electrochemical reactivity of Magnéli phase TiO₂ anodes[J]. J. Mater. Chem. A, 2018, 6: 23828-23839.

[91] Geng Z G, Kong X D, Chen W W, et al. Oxygen vacancies in ZnO nanosheets enhance CO_2 electrochemical reduction to CO[J]. Angew. Chem., Int. Ed., 2018, 57: 6054-6059.

[92] Nowotny J. Titanium dioxide-based semiconductors for solar-driven environmentally friendly applications: impact of point defects on performance[J]. Energy Environ. Sci., 2008, 1: 565-572.

[93] Qiu X Q, Miyauchi M, Yu H G, et al. Visible-light-driven Cu(Ⅱ)-$(Sr_{1-y}Na_y)(Ti_{1-x}Mo_x)O_3$ photocatalysts based on conduction band control and surface ion modification[J]. J. Am. Chem. Soc., 2010, 132: 15259-15267.

[94] Xie J F, Zhang H, Li S, et al. Defect-rich MoS_2 ultrathin nanosheets with additional active edge sites for enhanced electrocatalytic hydrogen evolution [J]. Adv. Mater., 2013, 25: 5807-5813.

[95] Liu M, Qiu X Q, Miyauchi M, et al. Cu(Ⅱ) oxide amorphous nanoclusters grafted Ti^{3+} self-doped TiO_2: an efficient visible light photocatalyst[J]. Chem. Mater., 2011, 23: 5282-5286.

[96] Liu M, Qiu X Q, Miyauchi M, et al. Energy-level matching of Fe(Ⅲ) ions grafted at surface and doped in bulk for efficient visible-light photocatalysts [J]. J. Am. Chem. Soc., 2013, 135: 10064-10072.

第 2 章

TiO_2 单晶电催化降解酚类污染物

2.1 概述

苯酚是一种重要的有机化工原料[1]。在石油精炼和化工生产等各种工业过程中,通常会不可避免地产生大量富含酚类物质的有毒废水,而它们往往难以进行生物降解[2]。电化学氧化法由于具有环境安全性、通用性和操作简便等优点,其研究受到越来越广泛的关注[3]。在电化学反应过程中,电极材料发挥着至关重要的作用[4-7],因而寻找或设计优异的电极材料、研究其在水处理过程中的表面性质和催化机理,具有重要的科学意义。

TiO_2 是一种典型的半导体材料,目前已被广泛应用于光化学水处理研究[2,8]。由于电导率低、阳极活性差[9-10],TiO_2 往往并不被认为是一种良好的电催化剂[10-11]。通过原子掺杂对 TiO_2 结构进行改性,TiO_2 的电催化性能可以大幅度提高[12-20];另外形貌、晶体结构和暴露晶面等参数也会对 TiO_2 的物理化学性质和催化性能产生很大影响,如 TiO_2 单晶(SCs)因内部晶体结构连续有序而具有较高的导电性。此外,高能{001}晶面表面原子的排列和协调决定了其独特的原子结构和电子结构,从而决定了其化学稳定性、吸附性能和催化反应活性。高能{001}晶面比低能{101}晶面具有更高的阳极催化活性,主要是由于低坐标表面原子的台阶密度、边缘和弯折密度较高,且有大量不饱和位点,因此 TiO_2 在经过形貌和晶面调控后可以成为优异的水处理阳极材料[21]。然而,它们的阳极氧化效率、反应特性和催化机理仍未被完全揭示。随着阳极偏压的增大,逐步加剧的析氧反应对污染物降解过程更为有效和实用。

在 TiO_2 电极表面,污染物通常发生电化学直接氧化或(水电解生成的)吸附态·OH介导的间接氧化[13-16,18-20]。在直接氧化过程中,除电子外没有其他物质参与反应,因此电流效率高、能耗低[3]。然而该过程存在一个主要的技术瓶颈,即污染物氧化生成的中间体聚合物会对电极造成严重污染[22]。在·OH介导的间接氧化过程中,污染物氧化效率高、反应速率快,因而更加实用,但析氧会导致能耗增高[1,23-25]。此外在·OH存在的情况下,污染物仍然可以进行直接氧化降解。因此,对阳极氧化过程的反应机理进行深入研究是非常有必要的。

阳极材料的氧化机理由其表面化学性质所决定。而·OH介导的间接氧化路径和直接氧化路径的比重,是决定阳极氧化效率和特征的重要因素[1,24-25]。例如,硼掺杂金刚石具有表面惰性,键合性与亲水性较差,生成的·OH键合作用较弱,具有高氧化过电位,对污染物的直接氧化能力较强[26-27];通过对PbO_2进行疏水树脂表面改性,其析氧过电位可显著增加,从而提高其电化学矿化效率[28]。许多报道认为,对于低能{101}晶面($0.44\ J\cdot m^{-2}$)暴露的锐钛矿TiO_2而言,即使有·OH生成,直接氧化过程也是污染物降解的主要途径[13-16,18-20]。相比之下,具有独特表面排列和配位的高能{001}暴露晶面($0.90\ J\cdot m^{-2}$)具有更高的原子台阶密度、低坐标表面原子的边缘和扭结,且有大量的不饱和键。这些性质影响着锐钛矿TiO_2的浸润性、氧键强度、催化活性和化学稳定性[29-30]。因此,深入了解{001}晶面在TiO_2电化学水处理中的影响和作用,具有非常重要的意义。

本章主要针对高能{001}暴露晶面TiO_2电催化降解酚类污染物的效率、特点和机理进行研究。首先,选取5种具有吸电子基或供电子基的对位取代酚作为电化学信号分子,探索其阳极氧化机理[1,24-25];随后,运用该电催化体系实现对不同种类实际废水的降解处理;最后,采用密度泛函理论(Density Functional Theory,DFT)对阳极氧化机理进行原子尺度的理论分析,旨在为钛基阳极材料的制备、改性和电化学水处理技术工艺的发展提供新的思路。

2.2 TiO_2 单晶电极的合成与电催化体系设计

2.2.1 TiO_2 单晶的制备与电极合成

我们采用水热法制备合成了TiO_2 SCs(表征及测试结果见图2.1、图2.2和图2.3)[22]。将9.0 mL(40 wt%)HF水溶液和6.0 mL蒸馏水缓慢加入到25 mL钛酸四丁酯(TBOT)中,在60 mL聚四氟乙烯高压反应釜中磁力搅拌0.5 h。

将微黄色前驱体在 180 ℃下恒温水热 24 h,冷却至室温后,将得到的白色粉末分别用乙醇、0.1 mol·L^{-1} NaOH 溶液和蒸馏水反复冲洗,以进行表面去氟,最后在 60 ℃条件下烘干备用。

选用德国 Degussa 公司的商用 TiO$_2$(即 P25,混合多晶,低能{101}晶面暴露,{101}暴露率>95%,平均粒径约为 25 nm,锐钛矿/金红石质量比=80∶20,BET 表面积约为 50 m^2·g^{-1})作为对照材料(图 2.3)。

采用已报道的方法[21],将 TiO$_2$(TiO$_2$ SCs 或 P25)沉积到经过预处理的钛片基底上,制备阳极电极。具体方法为:将质量分数为 3% 的 TiO$_2$ 粉末与聚乙二醇(PEG)的混合物加入到 100 mL 乙醇中,超声分散 30 min,再将得到的悬浊液通过滴涂法均匀涂抹在钛片基体上,120 ℃烘干,随后在 430 ℃、N$_2$ 气氛中煅烧 2.0 h,以完全去除有机黏合剂,使 TiO$_2$ 薄膜牢固地烧结到钛片基底上。

PbO$_2$/Sb-SnO$_2$/Ti(以下简写为 PbO$_2$/Ti)电极分为 3 层(图 2.3):钛片为底层,作为保护膜的热沉积 Sb-SnO$_2$ 为中间层,电沉积 PbO$_2$ 为最外层[27]。引入 PbO$_2$ 最外层与钛片基底支架之间的 Sb-SnO$_2$ 中间层,目的是防止钛片基底氧化失活,提高负载 PbO$_2$ 的电导率和机械强度。具体方法为:将饱和 Pb(NO$_3$)$_2$ 溶液与正丁醇混合后均匀分散在抛光的钛片上,然后在 80 ℃条件下加热 10 min,随后在 500 ℃条件下煅烧 10 min,这个过程重复 10 次,最后在 500 ℃条件下煅烧 60 min。电化学反应发生在电极/电解质的界面上,因此氧化效率只取决于外层电极材料的电催化活性。XRD 和 XPS 的测试结果证实了在制备完成的 PbO$_2$/Ti 电极表面没有任何 Sb-SnO$_2$ 组分暴露、参与电化学阳极氧化降解污染物,从而排除了 Sb-SnO$_2$ 中间层对 PbO$_2$ 最外层阳极氧化过程的可能贡献。

2.2.2
对位取代酚电催化降解及废水处理体系

在一个装有磁力搅拌器的圆柱形单室电解池中进行电化学降解实验。阳极为 TiO$_2$/Ti、P25/Ti 和 PbO$_2$/Ti 电极,阴极为钛片,阳极和阴极的有效电极面积均为 10.0 cm^2,电极间距为 1.0 cm。反应溶液体积为 500 mL,额外添加 0.1 mol·L^{-1} Na$_2$SO$_4$ 作为支持电解质,每种对位取代酚的浓度均为 300 mg·L^{-1}。实际水样包括巢湖地表水、合肥市政污水和工厂印染废水。实际水样均使用 0.45 μm 滤膜进行过滤,并用等体积的 0.2 mol·L^{-1} Na$_2$SO$_4$ 溶液进行稀释,以保证足够的离子强度和电导率(0.1 mol·L^{-1} Na$_2$SO$_4$)。电流密度分别控制为

25 mA·cm^{-2}、50 mA·cm^{-2} 和 7 mA·cm^{-2}。反应时间为 9.0 h，每间隔 60 min 进行取样分析，所有实验平行 3 次，计算出平均结果。

2.2.3
TiO$_2$ 单晶的材料表征与测试分析

通过场发射扫描电镜（FE-SEM，SIRION200，FEI Co.，the Netherlands）和高分辨率透射电镜（HRTEM，JEM-2100，JEOL Co.，Japan）对样品进行形貌表征。通过 X 射线衍射（XRD，X'Pert，PANalytical BV，the Netherlands）分析材料的晶体结构。通过 X 射线光电子能谱（XPS，PHI 5600，Perkin-Elmer Inc.，USA）对化学成分和价带光谱进行表征。使用溴化钾压片技术，通过傅里叶变换红外光谱仪（FTIR，Magna-IR 750，Nicolet Instrument Co.，USA）在 4000～400 cm^{-1} 范围内记录红外光谱。通过水接触角分析仪（JC2000A，Powereach Co.，Shanghai，China）测量材料的亲水性。通过电化学析氧反应（OER）和 LSV 对 0.1 mol·L^{-1} 含或不含对位取代酚的 Na$_2$SO$_4$ 水溶液进行扫描测试，扫描速度为 5～100 mV·s^{-1}。

在·OH 半定量捕获荧光（PL）测试中，TiO$_2$/Ti、P25/Ti 和 PbO$_2$/Ti 的电极面积为 10.0 cm^2，反应溶液为 80 mL 的 3.0 mmol·L^{-1} 对苯二甲酸、0.01 mol·L^{-1} NaOH 和 0.1 mol·L^{-1} Na$_2$SO$_4$ 水溶液，转速为 500 r·min^{-1}，每 5 min 进行取样分析一次。

ESR 捕获·OH 测试以 5,5-二甲基-1-吡咯烷酮 n-氧化物（DMPO，100 mmol·L^{-1}）为自旋捕获剂，通过电子顺磁共振（ESR）（A300，Bruker Co.，Germany）检测反应产物。

通过高效液相色谱法（HPLC-1100，Agilent Inc.，USA），用 Hypersil-ODS 反相柱和 VWD 检测器对取代苯酚、对硝基苯酚（p-NO$_2$）、对羟基苯甲醛（p-CHO）、苯酚（p-H）、对甲酚（p-CH$_3$）和对甲氧基苯酚（p-OCH$_3$）进行测定。流动相为水和甲醇混合物（体积比＝50∶50），流速为 1.0 mL·min^{-1}。紫外检测器波长设置为：314 nm 处为 p-NO$_2$，270 nm 处为其他对位取代酚。通过测定总有机碳（TOC）考察降解矿化效率，由 TOC 分析仪（Vario TOC cube，Elementar Co.，Germany）测试计算。通过荧光分光光度计（Fluorescent Spectrometer，FL1008M018，Cary Eclipse Co.，USA）检测对苯二甲酸的·OH 捕获 PL 信号。

2.2.4
TiO$_2$ 单晶修饰电极制备与电化学表征

将原始空白的玻碳电极(GCE,直径为 3.0 mm)通过 0.3 μm 和 0.05 μm 的氧化铝粉末进行机械抛光打磨,在蒸馏水中进行超声清洗。然后,在 0.5 mol·L^{-1} H$_2$SO$_4$ 中以 100 mV·s^{-1} 扫速,在 $-1.0\sim1.0$ V 电压范围下通过 CV 法进行电极活化处理。将 2.0 mg 催化剂超声分散在 2.0 mL 异丙醇中,滴加 4.0 μL 分散液到抛光活化后的玻碳电极表面,自然风干,完成电极修饰。

2.2.5
阳极氧化峰电位

对于吸附控制的完全不可逆电极过程,E_p 定义为

$$E_p = E_0 + \left(2.303\frac{RT}{\alpha nF}\right)\lg\frac{RTk_0}{\alpha nF} + \left(\frac{2.303RT}{\alpha nF}\right)\lg v \quad (2.1)$$

式中,α 为传递系数(通常被认为是 0.5),k_0 为标准反应速率常数,n 为电子转移数,v 为扫速,E_0 为标准氧化还原电位,R 为气体常数,T 为绝对温度,F 为法拉第常数。

5 种对位取代酚在 TiO$_2$ SCs 材料表面发生氧化反应的电子转移数约为 1,这与参考文献一致。在 $0.0\sim1.0$ V 的单周期 LSV 扫描测试中,以饱和甘汞电极(SCE)为参比电极,每种对位取代酚仅在 TiO$_2$/Ti 电极上释放 1 个电子进行电化学转化(式(2.1)、图 2.5 和图 2.9),未发生矿化;只有在阳极氧化过程中,才能完全转化和部分矿化(图 2.3)。因此,将电子转移数 n 设置为 1。

2.2.6
污染物扩散系数

采用计时电流法测定了 5 种对取代苯酚水溶液在改性玻碳电极表面的扩散

系数(图 2.3 和图 2.4)。电流(I)与 $t^{-1/2}$ 在不同 5 种对位取代酚浓度下的曲线呈现出不同斜率的直线(图 2.3 和图 2.4)。根据得到的斜率,可以用 Cottrell 方程计算出特定的扩散系数:

$$I = \frac{nFA\,CD^{\frac{1}{2}}}{\pi^{\frac{1}{2}}\,t^{\frac{1}{2}}} \tag{2.2}$$

式中,I 为感应电流,n 为电子的数目($n=1$),F 为法拉第常数($F=96485\,\text{C}\cdot\text{mol}^{-1}$),$A$ 为工作电极几何面积($0.196\,\text{cm}^2$),D 为扩散系数($\text{cm}^2\cdot\text{s}^{-1}$),$C$ 为溶液中对位取代酚的浓度($\text{mol}\cdot\text{cm}^{-3}$),$t$ 为扫描时间(s)。

2.2.7

TiO_2/Ti 电极表面初始污染物浓度

初始表面浓度($\varGamma$)可以通过 TiO_2 SCs/GCE、P25/GCE 和 PbO_2/GCE 电极在不同扫速(v)(图 2.7~图 2.9)条件下的 LSV 测试谱图来计算。当扫速增加时,阳极氧化峰电位也相应升高,通过线性变化的峰电位(E_p)与 $\ln v$ 拟合,说明这个过程发生了不可逆的电化学阳极反应。

在吸附污染物发生不可逆电化学阳极反应时,氧化峰电流(i_p)可表示为[1]

$$i_p = \frac{n\alpha n_a F^2 Av\varGamma}{2.718RT} \tag{2.3}$$

其中,n 为电子转移数($n=1$),α 为电荷转移系数(通常被认为是 0.5),n_a 为电子交换的数量($n_a=1$),A 为有效电极面积($A=0.196\,\text{cm}^2$),F 为法拉第常数,R 为气体常数($R=8.3\,\text{J}\cdot\text{mol}^{-1}\cdot\text{K}^{-1}$),$T$ 为热力学温度($T=298\,\text{K}$),$\varGamma$ 为污染物的初始表面浓度($\text{mol}\cdot\text{cm}^{-2}$),$v$ 为扫速($\text{V}\cdot\text{s}^{-1}$)。

因此,初始表面浓度 $\varGamma$ 可以通过以下公式进行计算[1]:

$$\varGamma = \frac{2.718RT[Slope]}{n\alpha n_a AF^2} \tag{2.4}$$

2.2.8

极限电流密度与能耗

对于质量传质控制下的非均相过程,一般有机化合物或具有相似质量传质系数的混合物的最大产率可通过极限电流密度（$i_{\lim}$）表示：

$$i_{\lim} = n \times F \times D_m \times C_{org} \qquad (2.5)$$

或者

$$i_{\lim} = 4 \times F \times D_m \times COD \qquad (2.6)$$

式中，$i_{\lim}$为有机物氧化的极限电流密度（$A \cdot m^{-2}$），n为参与有机物矿化反应的电子数量，F为法拉第常数，D_m为质量传质系数（$m \cdot s^{-1}$），C_{org}为溶液中有机物的浓度（$mol \cdot m^{-3}$），COD为溶液中的化学需氧量（$mol \cdot m^{-3}$）。

在恒流电解时，施加的电流密度（$i_{appl.}$）确定了两种不同的工作状态：

（1）$i_{appl.} < i_{\lim}$：电解在电流控制下，阳极氧化过程中形成降解中间体，电流效率为100%，有机物/COD随反应时间线性下降。

（2）$i_{appl.} > i_{\lim}$：电解过程处于质量传质控制下，有机物完全矿化为CO_2，参与二次反应（如析氧或电解质分解），导致电流效率下降。在这种情况下，有机物/COD去除率呈指数趋势。

采用Cottrell方程计算了5种对位取代酚在传统三电极体系中不同浓度下的极限电流测量结果，从而得到它们在TiO_2 SCs/GCE上的质量传质系数（D_m，$m \cdot s^{-1}$，图2.4）。计算得到的扩散系数与它们在电解体系中的初始表面富集浓度呈正相关关系，进一步揭示了TiO_2阳极氧化对位取代酚的机理和特性（图2.10和图2.16）。

矿化电流效率（MCE，%）可由TOC去除值计算得到[3]

$$MCE = \frac{\Delta(TOC)_{exp}}{\Delta(TOC)_{theor}} \times 100\% \qquad (2.7)$$

$\Delta(TOC)_{exp}$为TOC的实际去除值，$\Delta(TOC)_{theor}$为TOC的理论去除值，假设电子电量$Q(=I \times t)$只在矿化反应消耗：

$$C_6H_5NO_3 + 12H_2O \rightarrow 6CO_2 + HNO_3 + 28H^+ + 28e^- \qquad (2.8.1)$$

$$C_7H_6O_2 + 12H_2O \rightarrow 7CO_2 + 30H^+ + 30e^- \qquad (2.8.2)$$

$$C_6H_5OH + 11H_2O \rightarrow 6CO_2 + 28H^+ + 28e^- \qquad (2.8.3)$$

$$C_7H_8O + 13H_2O \rightarrow 7CO_2 + 34H^+ + 34e^- \qquad (2.8.4)$$

$$C_7H_8O_2 + 12H_2O \rightarrow 7CO_2 + 32H^+ + 32e^- \qquad (2.8.5)$$

$$\Delta(TOC)_{theor} = \frac{\frac{I \times t}{n_e \times F} \times n_c \times M \times 10^3}{V} \quad (mg \cdot L^{-1}) \qquad (2.8.6)$$

式中，I为电流强度（A），t为降解时间（s），F为法拉第常数，n_e为电子转移数，n_c为有机化合物的碳原子数，M为碳的原子量（$M = 12.0 \text{ g} \cdot mol^{-1}$），$V$为样品溶液的体积（L），对硝基苯酚、对羟基苯甲醛、苯酚、对甲酚、对羟基苯甲醚的n_e

分别为 28、30、28、34 和 32，n_c 为 6 或 7。

平均电化学能量效率（E_c，kg·kWh^{-1}）定义为单位电化学能耗（kWh）去除 TOC(kg) 的平均量[3]：

$$E_c = \frac{(TOC_0 - TOC_t)V}{U_{cell} I \Delta t} \quad (2.9)$$

式中，$(TOC_0 - TOC_t)$ 为 TOC 去除值，V 为反应器体积（L），U_{cell} 为平均槽电压（V），I 为电流（A），Δt 为降解时间（h）。

2.2.9

电催化过程的 DFT 理论计算

本章采用 CASTEP 编码研究了发生在锐钛矿 TiO_2 晶体表面的不同种类吸附的几何结构[31]，通过广义梯度近似（Generalized Gradient Approximation，GGA）和 PBE（Perdew，Bueke and Ernzerhof）泛函进行全电子计算[32-33]。在能量小于 1×10^{-3} eV·cell^{-1} 时，得到所有结构的最小能量。在 Brillouin 区，本研究利用 750 eV 的平面波截止能量和 0.04 A^{-1} 的 Monkhorst-Pack k 点分离进行了计算。为了在速率和精度之间取得平衡，本研究选取了被广泛用于半导体、绝缘体和非磁性金属研究的快速质量设置。在这里，能量容限设为 1×10^{-3} eV·cell^{-1}。该设置比精细设置快一个数量级，因此所产生的结果可以足够准确地用于探索性研究。为了搜索过渡状态并确定活化能垒，本研究采用了线性同步轨道（LST）/二次同步轨道（QST）方法[34-35]。通过以下公式计算吸附在 TiO_2 表面物质的吸附能（ΔE_{ads}）：

$$\Delta E_{ads}(\alpha) = E(\alpha\text{-sur}) - E(\text{free-}\alpha) - E(\text{clean-sur}) \quad (2.10)$$

式中，$E(\text{clean-sur})$ 为 TiO_2 表面的自由能，$E(\text{free-}\alpha)$ 为被吸附物在真空下的自由能，$E(\alpha\text{-sur})$ 为 TiO_2 表面所有组分的总自由能。

2.3
TiO₂单晶电催化降解对位取代酚的效能与机理分析

2.3.1
TiO₂单晶的形态结构特征

所制备的高能{001}晶面暴露 TiO₂ SCs 的 SEM、TEM 照片如图 2.1(a)和(b)所示,其形貌为均一性良好的超薄纳米片,平均尺寸约为 80 nm,厚度约为 10 nm。通过测量发现,TiO₂ SCs 为切角八面体,由 8 个侧向等效面和 2 个水平等效面构成,上下的水平等效面均为高能{001}晶面,暴露比例约为 80%。图 2.1(c)为 TiO₂ SCs 的高分辨率透射照片,显示了{001}区的衍射点和{200}区 1.9×10^{-10} m 的原子晶格间距。所观察到的条纹与界面距离相对应,和锐钛矿 TiO₂{101}低折射率面的晶格间距相吻合[29-30]。此外,晶格条纹清晰可见,表明其结晶度较高。

利用 XRD 进一步表征了 TiO₂ SCs 的晶体结构(图 2.1(d))。衍射图谱表明,制备的样品为纯锐钛矿相(四方位相,I 41/amd,JCPDS 21-1272),衍射信号分别对应锐钛矿{101}、{004}和{200}晶面的 29°、44°、56°三个衍射峰[29-30]。此外,TiO₂ SCs 在锐钛矿相呈结晶状,衍射峰较窄且较尖,说明水热合成的材料具有较好的结晶性。这一结果与 HRTEM 图像高度一致(图 2.1(c))。

2.3.2
TiO₂/Ti 阳极氧化对位取代酚的效能评估

本小节研究了具有吸电子基或供电子基的对取代苯酚的电化学氧化反应。结果表明,在 25 mA·cm⁻² 时,p-OCH₃ 和 p-NO₂ 的降解速率明显快于 p-CH₃、p-CHO、p-H(图 2.3)。随着电流密度的增加,p-CH₃、p-CHO、p-H 的去除率有所提高,但仍低于 p-OCH₃ 和 p-NO₂(图 2.3)。与{101}晶面暴露的 P25 相比,

▷ 第2章

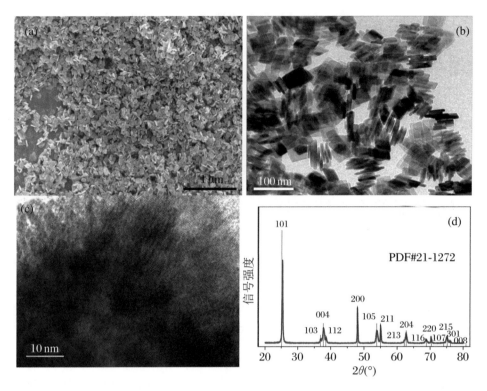

图 2.1　高能{001}晶面暴露 TiO_2 SCs 的 SEM(a)、TEM(b)、HRTEM(c)、XRD(d)表征

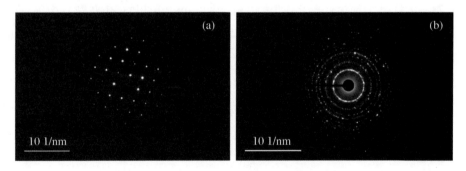

图 2.2　高能{001}晶面暴露 TiO_2 单晶(a)和低能{101}晶面暴露 P25 多晶(b)的选区电子衍射(SAED)图谱

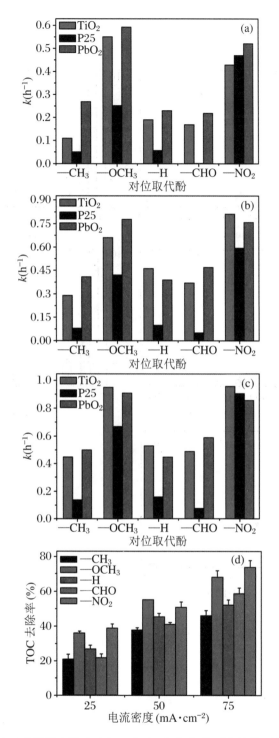

图 2.3 在 3 种不同电流密度下 TiO_2/Ti 阳极氧化对位取代酚的反应速率：(a) 25 mA·cm^{-2}；(b) 50 mA·cm^{-2}；(c) 75 mA·cm^{-2}；(d) TOC 的去除率

{001}晶面暴露的 TiO_2 SCs 具有更高的阳极氧化活性(图2.3、表2.1),但仍低于 PbO_2(图2.3、表2.1)[7]。

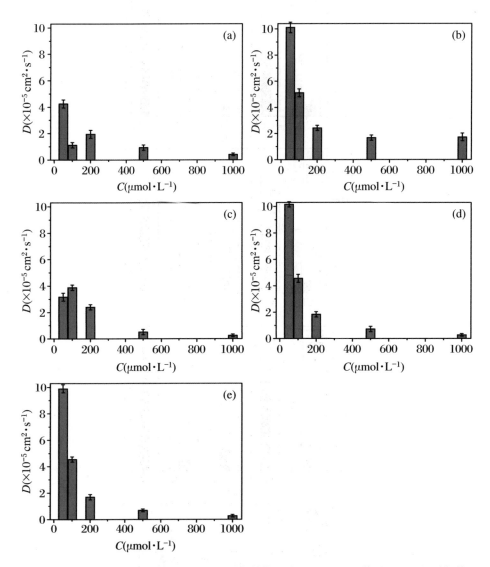

图 2.4　TiO_2/GCE 测定对甲酚(p-CH_3)(a)、对羟基苯甲醚(p-OCH_3)(b)、苯酚(p-H)(c)、对羟基苯甲醛(p-CHO)(d)和对硝基苯酚(p-NO_2)(e)分别在 $50.0\ \mu mol \cdot L^{-1}$、$100.0\ \mu mol \cdot L^{-1}$、$200.0\ \mu mol \cdot L^{-1}$、$500.0\ \mu mol \cdot L^{-1}$ 和 $1000.0\ \mu mol \cdot L^{-1}$ 浓度下的扩散系数 D

一些酚类物质由于表面富集较弱,降解不符合一级反应动力学方程(图2.5(a))。图2.3(d)为 TiO_2 SCs 的电化学矿化效果。$25\ mA \cdot cm^{-2}$ 时,p-OCH_3 和 p-NO_2 主要转化为可溶性中间产物[27-28,30]。p-OCH_3 和 p-NO_2 的 TOC 去除率分别为 ~37% 和 ~39%。尽管 p-CH_3、p-CHO、p-H 的 TOC 去除率较低,但仍可观察到它们的电化学转化(图2.1(d))。在 $75\ mA \cdot cm^{-2}$ 时,随着中间产物的减少,矿化率提高到 45%~75%。

表 2.1 TiO_2/Ti、P25/Ti 和 PbO_2/Ti 对 5 种对位取代酚的阳极氧化特性

对位取代酚[①]	电流密度 $(mA \cdot cm^{-2})$	阳极								
		TiO_2/Ti			P25/Ti			PbO_2/Ti		
		k (h^{-1})[②]	MCE[③]	E_c $(kg \cdot kWh^{-1})$[④]	k (h^{-1})[②]	MCE[③]	E_c $(kg \cdot kWh^{-1})$[④]	k (h^{-1})[②]	MCE[③]	E_c $(kg \cdot kWh^{-1})$[④]
—CH_3	25	0.11	9.83%	0.003	0.05	6.34%	0.002	0.27	11.64%	0.004
	50	0.29	8.89%	0.002	0.08	6.01%	0.001	0.41	10.84%	0.003
	75	0.45	7.18%	0.001	0.14	5.73%	0.001	0.50	9.39%	0.002
—OCH_3	25	0.55	14.69%	0.006	0.25	10.84%	0.004	0.59	16.47%	0.007
	50	0.66	11.22%	0.003	0.42	9.14%	0.002	0.78	14.68%	0.004
	75	0.95	9.25%	0.002	0.67	7.96%	0.002	0.91	13.29%	0.003
—H	25	0.19	12.42%	0.004	0.06	9.87%	0.003	0.23	15.53%	0.005
	50	0.46	10.35%	0.003	0.10	8.32%	0.002	0.39	14.35%	0.004
	75	0.53	7.97%	0.002	0.16	7.64%	0.001	0.45	11.75%	0.002
—CHO	25	0.17	9.06%	0.004	0.00	6.98%	0.002	0.22	10.87%	0.004
	50	0.37	8.45%	0.002	0.05	6.11%	0.001	0.47	9.98%	0.003
	75	0.49	8.10%	0.002	0.08	4.39%	0.001	0.59	7.69%	0.002

续表

对位取代酚	电流密度 (mA·cm^{-2})	阳极 TiO$_2$/Ti			阳极 P25/Ti			阳极 PbO$_2$/Ti		
		k (h^{-1})	MCE	E_c (kg·kWh^{-1})	k (h^{-1})	MCE	E_c (kg·kWh^{-1})	k (h^{-1})	MCE	E_c (kg·kWh^{-1})
—NO$_2$	25	0.43	12.17%	0.006	0.47	9.82%	0.004	0.52	15.75%	0.008
	50	0.81	7.96%	0.003	0.59	6.98%	0.002	0.76	14.39%	0.005
	75	0.96	7.70%	0.002	0.91	5.23%	0.001	0.86	13.28%	0.004

注：① 污染物浓度=300 mg·L^{-1}，溶液体积=500 mL，不调控 pH，搅拌转速为 500 r·min^{-1}，反应时间=9.0 h；

② $\ln(C_0/C_t) = kt$，C_0 为对位取代酚在 $t=0$ 时刻的浓度，C_t 为对位取代酚在 $t=t$ 时刻的浓度；

③ $MCE = \dfrac{\Delta(TOC)_{exp}}{\Delta(TOC)_{theory}} \times 100\%$，其中 $\Delta(TOC)_{exp}$ 为 TOC 去除的实验值，$\Delta(TOC)_{theory}$ 为 TOC 去除的理论值；

④ $E_c = \dfrac{(TOC_0 - TOC_t)V}{U_{cell} \int I \Delta t}$，其中 $(TOC_0 - TOC_t)$ 为降解前后的 TOC 下降值，V 为反应溶液体积(L)，U_{cell} 为电解过程中的平均电压值(V)，I 为电流值(A)，Δt 为电解时间(h)。

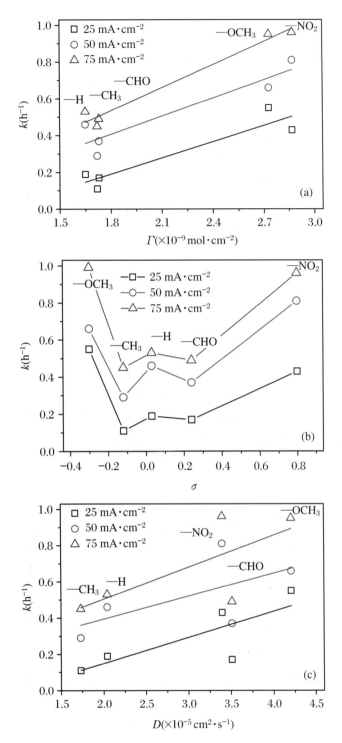

图 2.5　对位取代酚的阳极氧化降解速率(反应速率常数 k)和每种酚对应的表面浓度 Γ(a)、Hammett 常数 σ(b)与扩散系数 D(c)之间的关系

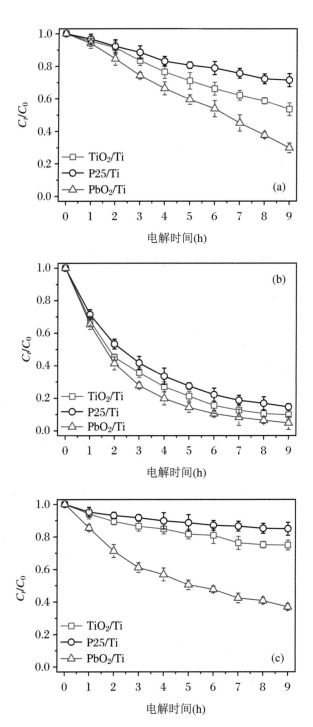

图 2.6 TiO$_2$/Ti、P25/Ti 和 PbO$_2$/Ti 对 5 种对位取代酚的降解(25 mA·cm^{-2}):p-CH$_3$(a)、p-OCH$_3$(b)、p-H(c)、p-CHO(d)、p-NO$_2$(e)和反应动力学比较(f)

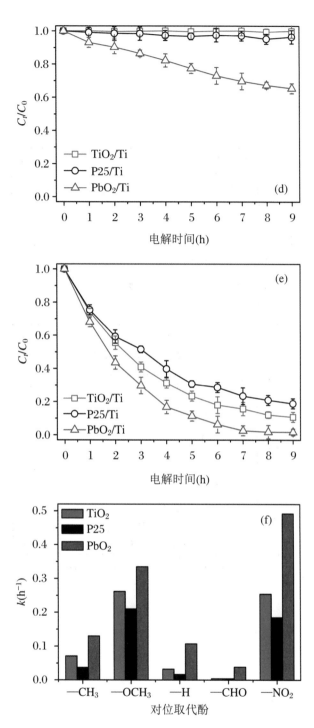

续图 2.6 TiO_2/Ti、$P25/Ti$ 和 PbO_2/Ti 对 5 种对位取代酚的降解($25\ mA \cdot cm^{-2}$):p-CH_3(a)、p-OCH_3(b)、p-H(c)、p-CHO(d)、p-NO_2(e)和反应动力学比较(f)

对 5 种对位取代酚在阳极氧化过程中的初始表面浓度的酚类(Γ)、Hammett 常数(σ)、扩散系数(D)与阳极活性(一级反应动力学常数 k)之间的关系进行了分析,其结果如图 2.4~图 2.6、图 2.9 和表 2.1 所示[1,24-25]。与参考文献相似(图 2.3 和图 2.7),TiO_2 SCs 的阳极反应动力学常数(k)与初始表面浓度(Γ)(图 2.5(a)和图 2.9)高度正相关。

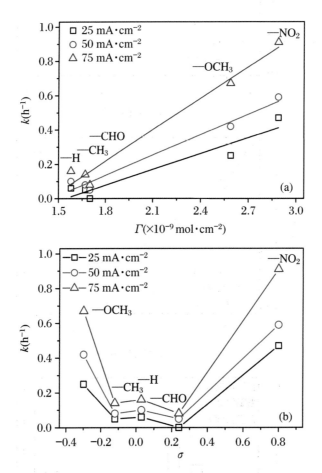

图 2.7　P25/Ti 降解 5 种对位取代酚过程的反应速率常数 k 与初始表面浓度 Γ(a)和 Hammett 常数 σ(b)的关系:电流密度分别为 25 mA·cm^{-2}、50 mA·cm^{-2} 和 75 mA·cm^{-2}

例如,在 50 mA·cm^{-2} 条件下,p-H、p-CH_3、p-CHO、p-OCH_3、p-NO_2 的反应速率常数(k)分别为 0.46 h^{-1}、0.29 h^{-1}、0.37 h^{-1}、0.66 h^{-1} 和 0.81 h^{-1},而对应的初始表面浓度(Γ)分别为 1.65×10^{-9} mol·cm^{-2}、1.72×10^{-9} mol·cm^{-2}、1.73×10^{-9} mol·cm^{-2}、2.73×10^{-9} mol·cm^{-2} 和 2.87×10^{-9} mol·cm^{-2}(图 2.9)。低富集浓度的 p-H 的反应速率常数(k)较高,为 0.46 h^{-1},可能与其

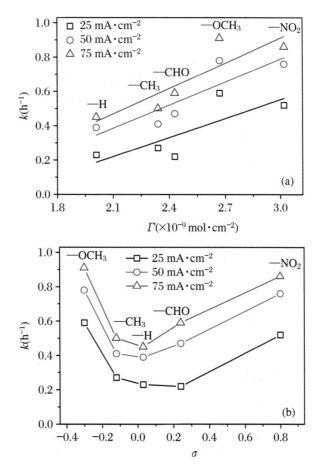

图 2.8　PbO_2/Ti 降解 5 种对位取代酚过程中反应速率常数 k 与初始表面浓度 Γ(a)和 Hammett 常数 σ(b)的关系：电流密度分别为 25 mA·cm^{-2}、50 mA·cm^{-2} 和 75 mA·cm^{-2}

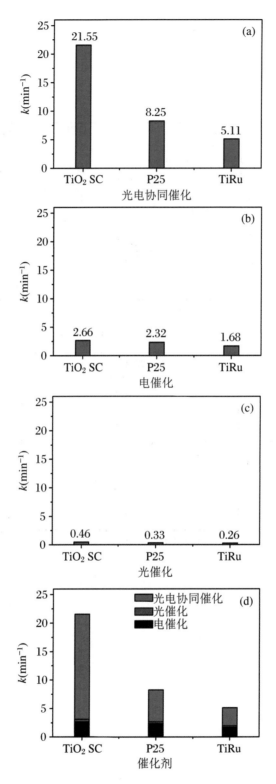

图 2.9　5 种对位取代酚在 LSV 扫描测试中扫速与阳极峰电流的关系以及它们在 TiO_2/GCE 上的初始表面浓度

对位取代基具有较高的吸电子性能有关[20]。在 25 mA·cm^{-2} 和 75 mA·cm^{-2} 时也观察到这种正相关趋势(图 2.5(a)),而 k 和 σ 没有任何线性相关性(图 2.5(b))。这些结果表明,对位取代酚的阳极氧化主要发生在 TiO_2 表面,受溶质向电极表面移动的传质控制[1,25]。通常来说,与供电子基相比,具有吸电子基的对位取代酚更难降解。其中,电化学反应活性在均相条件下与 Hammett 常数无关(图 2.5(b)),而在非均相条件下与污染物表面富集浓度关系较强(图 2.5(a))。根据不同扫速下的 LSV 测量,晶面形貌调控的锐钛矿{001}TiO_2 的阳极特性与 PbO_2 相似(图 2.1)。

在此基础上,提出表面结合态·OH 介导的阳极氧化机理[25]。对位取代酚降解过程中阳极活性(k)与扩散系数(D)之间的正相关关系为这一机理提供了直接证据(图 2.5c)。这种非均相特征主要归结于低电压下电极表面结合物·OH(·OH_{bound}, $E^0 = 1.23$ V/SHE)的局域结构形态,它是水分解和析氧过程的主要中间产物。在此过程中形成吸附在高能{001}晶面 TiO_2 表面的吸附态·OH(表 2.2 和图 2.1),而不是类似于 BDD 电极的游离态·OH(·OH_{free}, $E^0 = 2.72$ V/SHE)[1,25]。游离态·OH 的催化反应活性与污染物的 Hammett 常数有关[24-25],考虑到表面介导·OH 的高活性热力学性质[3],则吸附态·OH 只能在非均相条件下攻击预吸附的污染物,且在较大程度上受限于传质。同时,在高电流密度下,阳极水原位分解所生成的氧气也可能氧化污染物(图 2.3 和表 2.1)。5 种典型对位取代酚的 Hammett 常数(σ)与反应动力学常数(k)之间不存在内在关联(图 2.5(b)),形成了类似火山形的曲线,这主要归结于高能{001}晶面的 TiO_2/Ti 电极表面结合·OH 介导的电化学氧化机理(图 2.3)。

TiO_2 阳极氧化 5 种对位取代酚的极限电流密度(i_{lim})分别为 5.25 mA·cm^{-2}、10.48 mA·cm^{-2}、5.86 mA·cm^{-2}、8.33 mA·cm^{-2} 和 6.59 mA·cm^{-2}(图 2.4 和图 2.11)。由于施加的电流密度($i_{appl.}$)高于其极限电流密度(i_{lim}),因此苯酚的阳极降解处于质量传质控制下,具有良好的稳定性。这些结果与降解趋势和提出的阳极机理一致(图 2.5 和图 2.3)。MCE 和 E_c 是表征电化学水处理的重要参数[3]。与{101}-TiO_2 相比,{001}-TiO_2 具有更高的 MCE 和 E_c(表 2.1),由于析氧反应的发生,MCE 随着电流密度的增大而不断减小[4-5],且低于 PbO_2(表 2.1)。

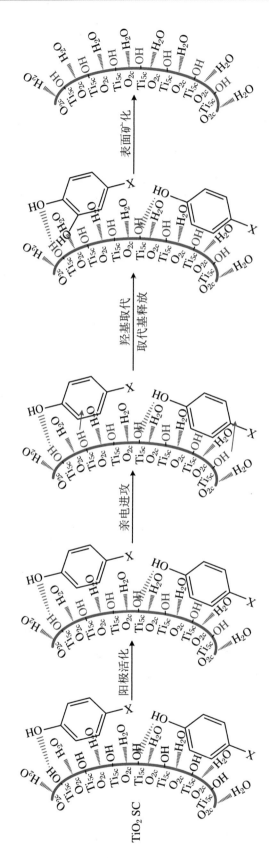

图2.10 高能{001}晶面暴露的TiO_2 SCs的阳极氧化机理研究

2.3.3
对位取代酚的阳极氧化机理

通过LSV扫描测试研究了直接氧化过程的贡献率(图2.11(a))。对位取代酚的阳极过电位(E_p)与Hammett常数(σ)呈正相关关系,各电位分别为:+0.94 V/SCE(p-H)、+0.72 V/SCE(p-CH$_3$)、+0.90 V/SCE(p-CHO)、+0.73 V/SCE(p-OCH$_3$)、+1.00 V/SCE(p-NO$_2$)(图2.11(b))。具有吸电子基和缺电子基的对位取代酚的直接氧化过程比具有供电子基和富电子环境的酚类更为困难[35]。它们在LSV中的反应速率常数(k)与E_p和I_p均不相关(图2.11(c)和2.11(d)),这些结果表明,在降解过程中,对位取代酚的电化学氧化不是通过直接氧化进行的(图2.3),而是通过·OH介导的机理进行的(图2.3)。

表面吸附·OH介导的阳极机理阐明了反应速率常数(k)和污染物初始表面浓度(Γ)以及Hammett常数(σ)之间的关系(图2.3和图2.5)[24-25]。这一机理主要归结于高能{001}晶面独特的表面化学性质,如超亲水性和通过表面—OH对水和·OH都有很强的亲和力(图2.12(a~c))(反应(2.11)~(2.12))[13-16,18-20]。FTIR(图2.12(b))证实了—OH与氧的强结合强度,使·OH与TiO$_2$ SCs严格结合,只攻击低析氧过电位条件下的预吸附污染物(反应(2.13)~(2.16))(图2.12(d))[25]。由于降解在恒流条件下进行,速率常数应由主要氧化剂·OH的选择性决定。

$$TiO_2 + H_2O \rightarrow TiO_2—H_2O_{ads} \rightarrow TiO_2—OH_{ads} + —H_{ads} \quad (2.11)$$

$$TiO_2—OH_{ads} + Bias \rightarrow TiO_2—·OH_{ads}^+ + e^- \quad (2.12)$$

$$TiO_2—·OH_{ads}^+ +\equiv C—H_{ads} + H_2O \rightarrow TiO_2—OH_{ads} +\equiv C—OH + H^+ + e^- \quad (2.13)$$

$$TiO_2—·OH_{ads}^+ +\equiv C—OH_{ads} + H_2O \rightarrow TiO_2—OH_{ads} + CO_2 + H_2O \quad (2.14)$$

$$TiO_2—·OH_{ads}^+ + H_2O_{ads} \rightarrow TiO_2—OH_{ads} + O_2\uparrow + 2H^+ + 2e^- \quad (2.15)$$

$$TiO_2—·OH_{ads}^+ + TiO_2—·OH_{ads}^+ \rightarrow TiO_2—H_2O_{2ads}$$
$$\rightarrow TiO_2—OH_{ads} + 1/2\ O_2\uparrow + H^+ \quad (2.16)$$

以对苯二甲酸为探针分子来测定·OH(图2.13(a))[36],TiO$_2$ SCs中生成的·OH浓度远高于P25对照,但低于PbO$_2$(图2.13(a))。由于对苯二甲酸不能通过直接电子转移进行羟基化反应,且被广泛用作探针来实现·OH的定量(图2.3)。因此,对苯二甲酸对羟基定量形成的络合物完全归结于·OH介导的反应路径[37]。TiO$_2$ SCs优异的·OH生成能力是由于它具有良好的阳极氧化

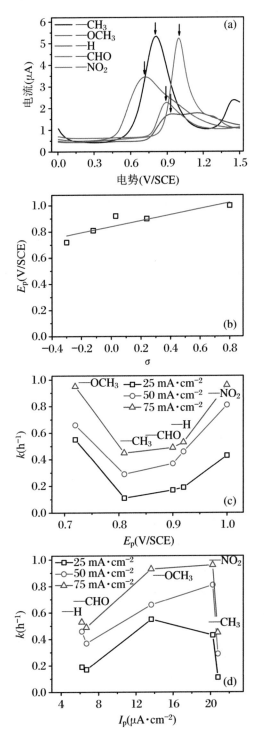

图 2.11 对位取代酚的 LSV 扫描谱图(a)、峰电位 E_p 和 Hammett 常数 σ 之间的关系(b)、反应速率常数 k 和峰电位 E_p 之间的关系(c)、反应速率常数 k 和峰电流 I_p 之间的关系(d)

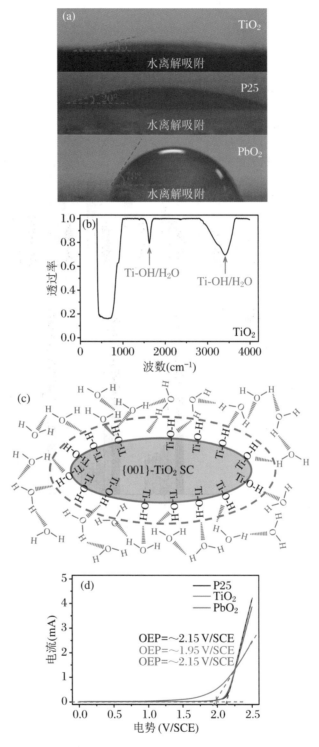

图 2.12 TiO_2、P25 和 PbO_2 的水接触角测试(a)、FTIR 测试(b)、水分子吸附模型(c)和电化学析氧电位测试(d)

活性(图2.3)[25]。虽然在光化学水处理中对{001}-TiO_2 具有较高的·OH生成能力已有较多的文献报道[17],但由于TiO_2活化机理和·OH形态结构的差异[22],其电化学水处理应用还是一个全新的研究领域。

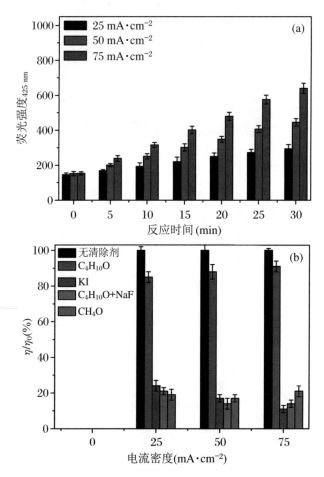

图2.13 TiO_2/Ti 电极阳极氧化过程中的·OH浓度测试(a)和抑制实验(b)

在光化学过程中,·OH的生成机理依赖于紫外光子($\lambda \leqslant 385$ nm)的价带活化(图2.16(a))。相比之下,电化学过程中的机理主要依赖于低阳极偏压时的导带活化($E < E_g = 3.2$ V),而在高阳极偏压时($E \geqslant E_g = 3.2$ V,图2.16(b))也可能发生价带活化。这是两个相对独立的·OH生成机理,因为在一般的导带和价带的电子结构以及宽的带隙3.2 eV中,主要由TiO_2的能带边缘的O 2p和Ti 3d构成。因此,在光化学水处理中,TiO_2构型(h_{VB}^+)的O 2p轨道上仍保留有价带通道,H_2O/OH^-氧化生成·OH(图2.16(a))。在电化学水处理中,·OH从H_2O/OH^-氧化生成TiO_2的导带通道出现在低偏压(h_{CB}^+)时的Ti 3d轨道,或高偏压(h_{CB}^+/h_{VB}^+)下的Ti 3d轨道和TiO_2的价带O 2p轨道。此外,

生成的·OH 在 TiO$_2$ 上的光化学和电化学机理完全不同。在光化学水处理中，由于价带通道在紫外光（UV）带隙激发下的氧化还原能力强，在溶液中产生高电位游离·OH（·OH$_{free}$，$E^0 = 2.72$ V/SHE）进行彻底的防污。在电化学水处理中，低电位表面吸附态·OH（·OH$_{bound}$，$E^0 = 1.23$ V/SHE）为部分生成转换在电极表面的、由于低氧化还原的驱动力下阳极的非本征带隙所激发，在其表面介导的催化机理中发挥主导作用（图2.10）。

为了进一步探索 TiO$_2$/Ti 电极的阳极氧化机理，分别使用了3种性质不同的·OH 清除剂进行实验。疏水型正丁醇由于对亲水 TiO$_2$ 表面的分子亲和力较低，主要抑制扩散层中的游离态·OH（图2.3、图2.12）[27,29]，因而 PNP 的降解率下降不到20%（图2.13(b)）。然而，在水溶液中加入 NaF 去除表面吸附自由基后，正丁醇的抑制作用明显增强。在 25 mA·cm^{-2}、50 mA·cm^{-2} 和 75 mA·cm^{-2} 时，PNP 氧化的一级反应速率常数从 0.43 h^{-1}、0.81 h^{-1} 和 0.96 h^{-1} 急剧下降到 0.09 h^{-1}、0.11 h^{-1} 和 0.13 h^{-1}。它们的抑制率分别为 4.8、7.4 和 7.4，与不含 NaF 的 1.2、1.1 和 1.1 相比，NaF 对照组的抑制率要高得多（图2.13(b)）。

此外，KI 通常作为空穴和吸附物·OH（表面介导的反应物种）的清除剂[25,38]，并将 PNP 的降解率降低了80%以上（图2.13(b)）。KI 和疏水丁醇对 NaF 的抑制作用与亲水性甲醇相当，亲水性甲醇是固体相和水相中最典型的·OH 清除剂（$k = 9.7 \times 10^8$ mol·L^{-1}·s^{-1}）。ESR 测定进一步证实了 PNP 降解的抑制作用。这些结果清晰地表明，表面吸附态·OH 在 TiO$_2$/Ti 电极上对取代苯酚的氧化过程中发挥着主导作用（图2.3和图2.13），而直接氧化路径即使在·OH 生成和水分解的情况下也可能只起次要的作用（图2.11）。

2.3.4
{001}晶面在阳极氧化过程中的作用

在电化学水处理中，H$_2$O 与 TiO$_2$ 的表面相互作用在·OH 生成和污染物降解过程中起主导作用[39-40]。其核心问题在于 H$_2$O 是否在 TiO$_2$ 表面发生分解，因为 H$_2$O 的分解可以同时引入—OH 和—H，并对 TiO$_2$ 的表面化学性质产生显著影响，有利于·OH 的生成和污染物的降解[39]。表面原子的配位数及其相互距离严重影响 H$_2$O/TiO$_2$ 的相互作用，不饱和 Ti 和 O 位点通过 Ti—O 键和 H—O 键对 TiO$_2$ 吸附水分子起主导作用[39]。因此，研究和了解 TiO$_2$ 表面吸

附的 H_2O 结构及其与表面—OH 的解离过程至关重要[40]。

与锐钛矿 TiO_2 的低能{101}晶面相比(50% Ti 3c + 50% Ti 2c + 50% O 3c + 50% O 2c, 0.44 J·m^{-2})，高能{001}晶面上的 Ti 和 O 原子处于完全不饱和状态(100% Ti 2c + 100% O 2c, 0.90 J·m^{-2})[29-30]。在能量转移和再分配过程中，吸附的 H_2O 受氧化物晶格静电场引导而重定向，活化和脱质子生成表面—OH（图 2.14 和图 2.15）。在{001}晶面上水的吸附能 $\Delta E_{ads}(H_2O)$ 变得更负、更稳定（表 2.2），这与红外光谱的结果（图 2.12(b)）一致。TiO_2 发生水分子吸附（图 2.14(a)）后，将一个 H 转移到附近的 O_{2c} 表面形成端部—OH（反应(2.17)、图 2.14(b、d)），或将一个 H 顶出与另一游离 H_2O 生成 H_3O^+（反应(2.18)、图 2.14(f、h)）：

$$TiO_2-H_2O \rightarrow HO-TiO_2-H \quad (2.17)$$

$$TiO_2-H_2O + H_2O \rightarrow TiO_2-OH + H_3O^+ + e^- \quad (2.18)$$

通过整体负能量变化(反应(2.17)($\Delta E = -1.330$ eV)、表 2.2)观察到，锐钛矿{001}晶面表面的水分解生成表面—OH 的过程（图 2.15），由于总能量为正，水的分解并不有利（$\Delta E_{Eq.(2.17)}$ 在{101}晶面以及 $\Delta E_{Eq.(2.19)}$ 在{101}和{001}晶面上，见表 2.2）。此外，在{001}晶面上，H_2O 分解（反应(2.17)～(2.18)）的活化能(E_a)的降低意味着表面—OH 形成的反应速率更高。从 H_2O 吸附结构的能量差异可以看出，被释放的 H 更倾向于与附近的 O 2c 结合，导致表面—OH 数量增多（图 2.15、表 2.2）。

在阳极极化下，表面—OH 失去一个电子所生成的表面介导·OH（反应(2.19)），是阳极污染物降解的主要活性物种（反应(2.13)和(2.14)）[3]。

$$TiO_2-OH_{ads} + Bias \rightarrow TiO_2-·OH_{ads}^+ + e^- \quad (2.19)$$

表 2.2 {001}/{101}-TiO_2 阳极氧化过程中的水分子吸附能($\Delta E_{ads}(H_2O)$)、羟基自由基吸附能($\Delta E_{ads}(·OH)$)、能量变化(ΔE_{Eq})和激发能(E_a)

晶面	$\Delta E_{ads}(H_2O)$ (eV)*	$\Delta E_{ads}(·OH)$ (eV)*	$\Delta E_{Eq.(2.17)}$ (eV)	$\Delta E_{Eq.(2.18)}$ (eV)	$E_{a,Eq.(2.17)}$ (eV)	$E_{a,Eq.(2.18)}$ (eV)	$\Delta E_{Eq.(2.19)}$ (eV)
{001}	-1.079	-17.110	-1.330	1.089	-0.010	3.831	-0.350
{101}	-0.719	-14.126	0.310	3.309	1.291	15.375	0.913

* TiO_2 表面吸附物质的吸附能通过以下公式计算：$\Delta E_{ads}(\alpha) = E(\alpha\text{-sur}) - E(\text{free-}\alpha) - E(\text{clean-sur})$，其中 $E(\text{clean-sur})$ 为 TiO_2 的表面自由能，$E(\text{free-}\alpha)$ 为吸附物质在真空状态下的自由能，$E(\alpha\text{-sur})$ 为 TiO_2 表面所有吸附组分的总自由能。

从{001}和{101}晶面（反应(2.19)）能量变化 -0.350 eV + 0.913 eV（表 2.2）可以看出，·OH 吸附到{001}晶面需要消耗更多的能量。这些结果与

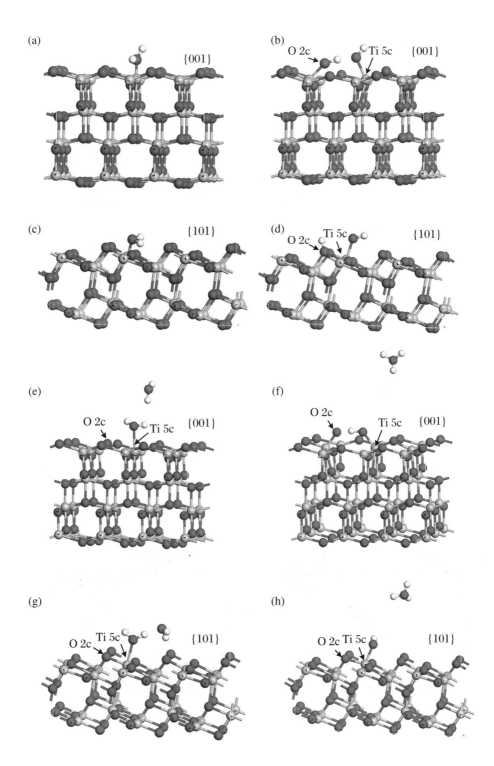

图 2.14 {001}(a、b、e 和 f)和{101}(c、d、g 和 h)的锐钛矿 TiO_2 表面吸附解离的优化计算结构:Ti 原子为浅灰色,O 原子为红色,H 原子为白色

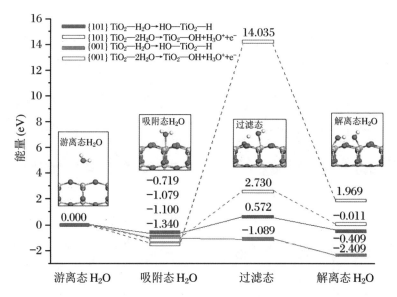

图 2.15 阳极氧化过程中锐钛矿 TiO_2 的高能{001}晶面和低能{101}晶面表面的水吸附和分解的能量谱

·OH 的测量值高度一致(图 2.13(a))。在{101}/{001}晶面上的模拟吸附能量计算值分别为 -17.110 eV 和 -14.126 eV,生成的 ·OH 也表现出对{001}晶面更高的亲和力(ΔE_{ads}(·OH),表 2.2)。DFT 计算的结果表明,以 Ti 5c 为中心的高能{001}晶面不仅有利于 ·OH 的生成,这种原子尺度的研究可以通过晶面调控工程为提高阳极材料的电化学水处理能力提供重要的信息。

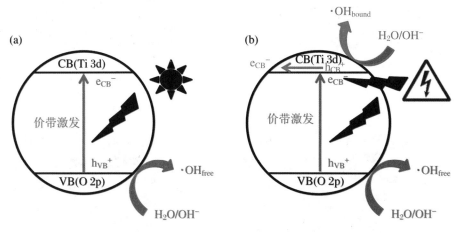

图 2.16 TiO_2 SCs 在 UV 紫外光照射(a)和施加阳极偏压(b)条件下降解污染物的激发过程示意图

2.3.5
实际废水的处理效果

使用所合成的 TiO_2 电极来降解实际废水,验证电化学技术处理含酚废水的可行性(图 2.17)。在 25 mA·cm^{-2} 与 9.0 h 降解条件下,计算出在{001}-TiO_2 电极上降解实际的酚类废水、城市污水、印染废水和地表水的一级反应速率常数分别为 0.173 h^{-1}、0.306 h^{-1}、0.207 h^{-1}、0.461 h^{-1} 和 0.24 h^{-1}。相比之下,P25 对照组的值分别为 0.092 h^{-1}、0.246 h^{-1}、0.129 h^{-1}、0.360 h^{-1} 和 0.200 h^{-1}(图 2.17(f))。与 P25 相比,{001}-TiO_2 具备较高的污染物去除效率和更快的反应动力学常数(稍低于 PbO_2)(图 2.17)。

本研究所采用的晶面调控方法显著提高了 TiO_2 对实际含酚废水的电化学降解活性。这种电导率和阳极活性的优势应主要归功于催化剂材料中的单晶结构和表面上共暴露的{101}/{001}晶面。由于晶界和界面状态较少,原子和电子结构连续有序的单晶结构非常有利于在体相内方便快速地转移电荷[21,26,29]。此外,共暴露的{101}/{001}晶面可以通过 ii 型带隙构型在内部建立热力学外延晶面结,允许在固体表面不同暴露晶面之间建立内置电场,有效地进行载流子分离,从而显示出增强的量子效率和催化性能[41]。与典型的工业尺寸稳定阳极(如 PbO_2)相比,高能{001}晶面暴露的 TiO_2 具有稳定性高、成本低、无毒、电化学水处理安全等优点。这些优点使其成为一种很有前途的电化学水处理阳极材料。

本章小结

本章工作充分利用晶面调控策略,以 TiO_2 为出发点,合成了具有高能{001}极性暴露晶面的单晶电极材料。选取 5 种含吸电子和供电子基的对位取代酚作为目标污染物,研究了 TiO_2 单晶对典型酚类污染物的电化学氧化效能。研究发现,含有供电子和吸电子基的 5 种典型对位取代酚的阳极氧化主要发生在 TiO_2 表面,且受污染物传质控制,氧化效率随着 Hammett 常数的增加而降低,而降解

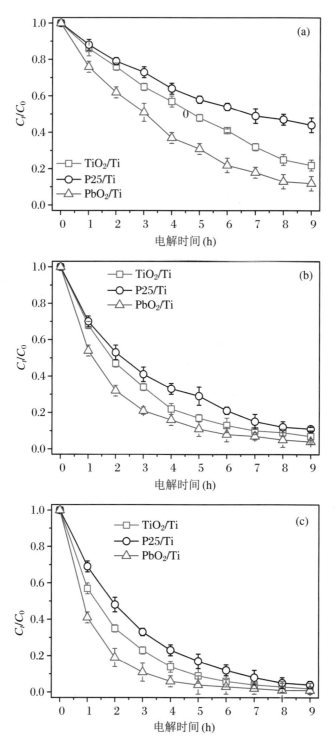

图 2.17 TiO_2/Ti、$P25/Ti$ 和 PbO_2/Ti 电极对实际酚类废水的电化学降解处理：混合酚溶液(a, COD = 500 mg·L^{-1})、二沉池出水(b, TOC = 7.0 mg·L^{-1})、印染废水 1(c, COD = 335 mg·L^{-1})、印染废水 2(d, COD = 335 mg·L^{-1})、地表水(e, COD = 59.0 mg·L^{-1}) 和动力学分析对比(f)

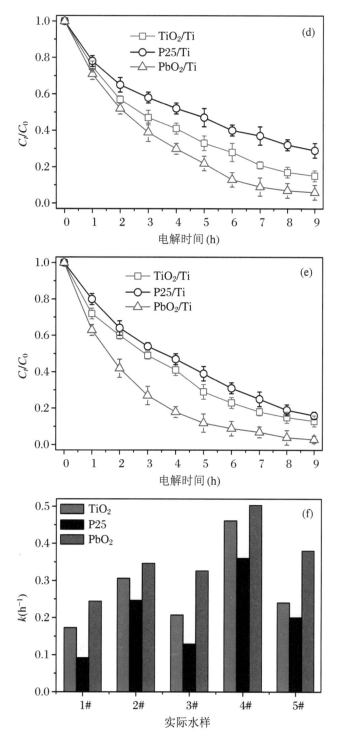

续图2.17 TiO_2/Ti、$P25/Ti$ 和 PbO_2/Ti 电极对实际酚类废水的电化学降解处理：混合酚溶液(a,COD=500 mg·L^{-1})、二沉池出水(b,TOC=7.0 mg·L^{-1})、印染废水1(c,COD=335 mg·L^{-1})、印染废水2(d,COD=335 mg·L^{-1})、地表水(e,COD=59.0 mg·L^{-1})和动力学分析对比(f)

速率则随着初始表面浓度的升高而加快;TiO₂ 单晶的电催化降解主要通过吸附态·OH 氧化和直接电子转移完成,且对实际含酚废水也具有较好的处理效果;DFT 计算表明,高能{001}晶面和单晶结构在 TiO₂ 电极表面结合态·OH 介导的酚类污染物降解机理中起主导作用。基于实验和理论分析结果,本章提出了 TiO₂ 单晶电化学氧化酚类污染物的机理(图 2.18),为进一步制备和应用 TiO₂ 阳极材料进行电化学水处理提供了科学依据,也为含酚废水提供了一种有效的处理方法。

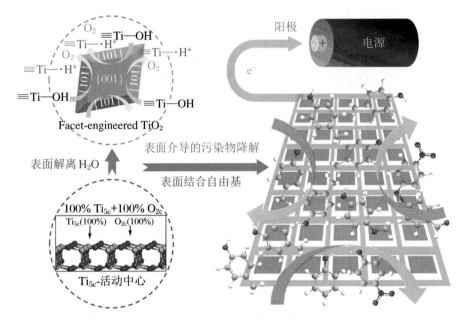

图 2.18 TiO₂ 电催化降解污染物机理示意图

参考文献

[1] Zhu X P, Shi S Y, Wei J J, et al. Electrochemical oxidation characteristics of p-substituted phenols using a boron-doped diamond electrode [J]. Environ. Sci. Technol., 2007, 41: 6541-6546.

[2] Awfa D, Ateia M, Fujii M, et al. Photodegradation of pharmaceuticals and personal care products in water treatment using carbonaceous-TiO₂ composites: a critical review of recent literature[J]. Water Res., 2018, 142: 26-45.

[3] Panizza M, Cerisola G. Direct and mediated anodic oxidation of organic

pollutants[J]. Chem. Rev., 2009, 109: 6541-6569.

[4] Martínez-Huitle C A, Rodrigo M A, Sirés I, et al. Single and coupled electrochemical processes and reactors for the abatement of organic water pollutants: a critical review[J]. Chem. Rev., 2014, 115: 13362-13407.

[5] Miklos D B, Remy C, Jekel M, et al. Evaluation of advanced oxidation processes for water and wastewater treatment: a critical review[J]. Water Res., 2018, 139: 118-131.

[6] Radjenovic J, Sedlak D L. Challenges and opportunities for electrochemical processes as next-generation technologies for the treatment of contaminated water[J]. Environ. Sci. Technol., 2015, 49: 11292-11302.

[7] Wu W Y, Huang Z H, Lim T T. Recent development of mixed metal oxide anodes for electrochemical oxidation of organic pollutants in water[J]. Appl. Catal., A, 2014, 480: 58-78.

[8] Chen X B, Mao S S. Titanium dioxide nanomaterials: synthesis, properties, modifications, and applications[J]. Chem. Rev., 2007, 107: 2891-2959.

[9] Li Y H, Liu P F, Pan L F, et al. Local atomic structure modulations activate metal oxide as electrocatalyst for hydrogen evolution in acidic water [J]. Nat. Commun., 2015, 6: 8064.

[10] Yang Y, Li J X, Wang H, et al. An electrocatalytic membrane reactor with self-cleaning function for industrial wastewater treatment[J]. Angew. Chem., Int. Ed., 2011, 50: 2148-2150.

[11] Yang Y, Wang H, Li J, et al. Novel functionalized nano-TiO_2 loading electrocatalytic membrane for oily wastewater treatment[J]. Environ. Sci. Technol., 2012, 46: 6815-6821.

[12] Chen D J, Chen C, Baiyee Z M, et al. Nonstoichiometric oxides as low-cost and highly-efficient oxygen reduction/evolution catalysts for low-temperature electrochemical devices[J]. Chem. Rev., 2015, 115: 9869-9921.

[13] Cho K, Hoffmann M R. Urea degradation by electrochemically generated reactive chlorine species: products and reaction pathways[J]. Environ. Sci. Technol., 2014, 48: 11504-11511.

[14] Cho K, Qu Y, Kwon D, et al. Effects of anodic potential and chloride ion on overall reactivity in electrochemical reactors designed for solar-powered wastewater treatment[J]. Environ. Sci. Technol., 2014, 48: 2377-2384.

[15] Kesselman J M, Weres O, Lewis N S, et al. Electrochemical production of hydroxyl radical at polycrystalline Nb-doped TiO$_2$ electrodes and estimation of the partitioning between hydroxyl radical and direct hole oxidation pathways[J]. J. Phys. Chem. B, 1997, 10: 2637-2643.

[16] Kim J, Kwon D, Kim K, et al. Electrochemical production of hydrogen coupled with the oxidation of arsenite[J]. Environ. Sci. Technol., 2014, 48: 2059-2066.

[17] Liu L, Chen X B. Titanium dioxide nanomaterials: self-structural modifications[J]. Chem. Rev., 2014, 114: 9890-9918.

[18] Park H, Vecitis C D, Choi W, et al. Solar-powered production of molecular hydrogen from water[J]. J. Phys. Chem. C, 2008, 112: 885-889.

[19] Park H, Vecitis C D, Hoffmann M R. Electrochemical water splitting coupled with organic compound oxidation: the role of active chlorine species [J]. J. Phys. Chem. C, 2009, 113: 7935-7945.

[20] Park H, Vecitis C D, Hoffmann M R. Solar-powered electrochemical oxidation of organic compounds coupled with the cathodic production of molecular hydrogen[J]. J. Phys. Chem. A, 2008, 112: 7616-7626.

[21] Zhang A Y, Long L L, Liu C, et al. Electrochemical degradation of refractory pollutants using TiO$_2$ single crystals exposed by high-energy {001} facets[J]. Water Res., 2014, 66: 273-282.

[22] Liu C, Zhang A Y, Si Y, et al. Photochemical anti-fouling approach for electrochemical pollutant degradation on facet-tailored TiO$_2$ single crystals [J]. Environ. Sci. Technol., 2017, 51: 11326-11335.

[23] Gao G D, Vecitis C D. Electrochemical carbon nanotube filter oxidative performance as a function of surface chemistry[J]. Environ. Sci. Technol., 2011, 45: 9726-9734.

[24] Torres R A, Torres W, Peringer P, et al. Electrochemical degradation of p-substituted phenols of industrial interest on Pt electrodes: attempt of a structure-reactivity relationship assessment[J]. Chemosphere, 2003, 50: 97-104.

[25] Zhu X P, Tong M P, Shi S Y, et al. Essential explanation of the strong mineralization performance of boron-doped diamond electrodes [J]. Environ. Sci. Technol., 2008, 42: 4914-4920.

[26] Lei Y Z, Zhao G H, Zhang Y G, et al. Highly efficient and mild

electrochemical incineration: mechanism and kinetic process of refractory aromatic hydrocarbon pollutants on superhydrophobic PbO_2 anode[J]. Environ. Sci. Technol., 2010, 44: 7921-7927.

[27] Zhao G H, Zhang Y G, Lei Y Z, et al. Fabrication and electrochemical treatment application of a novel lead dioxide anode with superhydrophobic surfaces, high oxygen evolution potential, and oxidation capability[J]. Environ. Sci. Technol., 2010, 44: 1754-1759.

[28] Zhou M H, Dai Q Z, Lei L C, et al. Long life modified lead dioxide anode for organic wastewater treatment: electrochemical characteristics and degradation mechanism[J]. Environ. Sci. Technol., 2005, 39: 363-370.

[29] Liu G, Yang H G, Pan J, et al. Titanium dioxide crystals with tailored facets[J]. Chem. Rev., 2014, 114: 9559-9612.

[30] Liu S G, Yu J G, Jaroniec M. Anatase TiO_2 with dominant high-energy {001} facets: synthesis, properties and applications[J]. Chem. Mater., 2011, 23: 4085-4093.

[31] Segall M D, Lindan P J D, Probert M J, et al. First-principles simulation: ideas, illustrations and the CASTEP code[J]. J. Phys.: Condens. Matter, 2002, 14: 2717-2744.

[32] Perdew J P, Burke K, Ernzerhof M. Generalized gradient approximation made simple[J]. Phys. Rev. Lett., 1996, 77: 3865-3868.

[33] Perdew J P, Chevary J A, Vosko S H, et al. Atoms, molecules, solids, and surfaces-applications of the generalized gradient approximation for exchange and correlation[J]. Phys. Rev. B, 1992, 46: 6671-6687.

[34] Govind N, Petersen M, Fitzgerald G, et al. A generalized synchronous transit method for transition state location[J]. Comp. Mater. Sci., 2003, 28: 250.

[35] Halgren T A, Lipscomb W N. The synchronous-transit method for determining reaction pathways and locating molecular transition states[J]. Chem. Phys. Lett., 1997, 49: 225-232.

[36] Qin Y X, Li G Y, Gao Y P, et al. Persistent free radicals in carbon-based materials on transformation of refractory organic contaminants (ROCs) in water: A critical review[J]. Water Res., 2018, 137: 130-143.

[37] Jing Y, Chaplin B P. Mechanistic study of the validity of using hydroxyl radical probes to characterize electrochemical advanced oxidation processes

[J]. Environ. Sci. Technol., 2017, 51: 2355-2365.

[38] Lai W W P, Hsu M H, Lin A Y C. The role of bicarbonate anions in methotrexate degradation via UV/TiO$_2$: Mechanisms, reactivity and increased toxicity[J]. Water Res., 2017, 112: 157-166.

[39] Sun C H, Liu L M, Selloni A, et al. Titania-water interactions: a review of theoretical studies[J]. J. Mater. Chem., 2010, 20: 10319-10334.

[40] Wang Z T, Wang Y G, Mu R T, et al. Probing equilibrium of molecular and desrotonated water on TiO$_2$(110)[J]. Proc. Natl. Acad. Sci., 2017, 114: 1801-1805.

[41] Zhang A Y, Wang W Y, Chen J J, et al. Epitaxial facet junctions on TiO$_2$ single crystals for efficient photocatalytic water splitting[J]. Energy Environ. Sci., 2018, 11: 1444-1448.

第 3 章

紫外光辅助 TiO$_2$ 单晶电催化降解污染物

3.1
概述

在水处理应用中,低偏压无析氧的电催化技术(Electrochemical Catalysis,EC)具有电流效率高、能耗低等优点[1]。污染物在电极/溶液界面吸附后以较低的速率发生直接氧化,除电子转移外不涉及任何其他反应。其主要瓶颈之一是污染物降解产生的中间体聚合物累积导致电极污染与失活[1]。光催化(Photochemical Catalysis,PC)作为一种高效且先进的水处理技术[2],可以为电催化中在阳极产生的中间体聚合物的去除提供有效策略[3-7]。当 PC 和 EC 耦合,形成光辅助电催化(Photo-assisted Electrochemical Catalysis,PEC)体系时,其偏压明显高于污染物和水的阳极氧化电位[8-9]。由此可以猜测存在如下协同效应:EC 中的阳极偏压和析氧反应促使 PC 活性提高,PC 中活性氧物质的生成促使 EC 活性的增强。

与普遍报道的氧析出后的间接 EC 和间接 PC 的高偏压反应相比[8-16],氧析出前的直接 EC 与间接 PC 低偏压反应具有电流效率高、能耗低和电极污染程度轻等优点,这对水处理应用具有更实用的价值。因此,控制 EC 的偏压,使其高于污染物的氧化电位、低于水的氧化电位,是一种可行的手段。然而目前还没有相关的研究报道。

在 PEC 反应中,电极材料需要同时具备较高的 PC 和 EC 活性。由于特定的晶体形态和电子结构,金属氧化物难以同时具备 EC 或 PC 活性[17],因而无法直接用于 PEC 反应。为了解决这个问题,将 EC 与 PC 催化剂在电极上进行复合,以期寻找制备 PEC 阳极材料的有效方法。通常将一种光催化剂(如 TiO_2 和 $\gamma\text{-}Bi_2MoO_6$)和另一种电催化剂(如 BDD、Pt 和过渡金属氧化物 RuO_2、SnO_2、PbO_2、IrO_2 和 Ta_2O_5)结合到同一电极上[8-16],来实现光、电双功能的复合过程,这也是目前大多数 PEC 体系的构建方法。该体系的主要瓶颈是表面反应活性位点的数量会因界面域的减少而受限[18]。近年来,$ZnWO_4$、Bi_2WO_6 和 $\gamma\text{-}Bi_2MoO_6$ 等单组分 PEC 电极被陆续开发出来[19-21],它们对污染物的协同降解作用优于 PC 和 EC 的总和。因此,开发简单、高效和稳定的阳极材料对 PEC 水处理应用具有重要意义。

TiO_2 具有良好的光催化活性[28-30],而高暴露{001}晶面的 TiO_2 SCs 已被证明是一种高效的阳极材料,在析氧前的低偏压条件下能够快速氧化降解污染物[22-27]。考虑到 TiO_2 SCs 的这种双功能性质,可以推测 PEC-TiO_2 体系很可能具备优良的水处理应用潜力。

本章工作提出一种新型的(紫外)光化学防污策略,即利用特异性晶面调控的 TiO_2 SCs 作为 PEC 催化剂,在低偏压析氧电位前对污染物进行阳极氧化[31]。以典型的环境内分泌干扰物双酚 A(BPA)作为目标污染物,以 EC 体系作为对照,从降解效率、防污性能和能耗等方面对 PEC 体系进行深入研究,同时考察 PEC 对垃圾渗滤液的处理效果,从而探索和验证光化学防污策略和基于 TiO_2 的 PEC 水处理体系的技术可行性。

3.2 TiO_2 单晶的紫外光-电性能与光电耦合催化体系设计

3.2.1 光电电极的制备与结构表征

利用水热法制备合成了具有优势高能{001}晶面的 TiO_2 SCs[24-25],具体操作同第 2 章。使用 P25 作为对照催化剂。将前驱体氯化物 $TiCl_4$、$RuCl_3$ 和 $SnCl_2$ 分别滴涂在抛光处理后的钛片表面进行高温煅烧,制备出性能稳定的 Sb-SnO_2 和 $Ti_{0.7}Ru_{0.3}O_2$ 电极。

将 TiO_2(TiO_2 SCs 或 P25)沉积在碳纸上制备阳极电极[24,34]。具体方法为:将质量分数为 3% 的 TiO_2 粉末与聚乙二醇(PEG,相对分子质量为 20000)的混合物加入到 100 mL 乙醇中,超声至少 30 min,将得到的悬浊液均匀地涂抹在 $4.0\ cm \times 1.5\ cm$ 的碳纸上;然后将样品在 120 ℃下烘干乙醇后,在马弗炉中加热至 430 ℃,恒温 2.0 h,完全除去有机黏结剂,将 TiO_2 牢固地烧结到碳纸基底上。该电极的 TiO_2 负载量为 0.30 mg(0.05 mg·cm^{-2}×6 cm^2)左右,对于反复进行的光电测量和 BPA 降解实验具有足够的稳定性。

通过场发射扫描电镜（FE-SEM，SIRION200，FEI Co.，the Netherlands）、高分辨透射电镜（HRTEM/SAED，JEM-2100，JEOL Co.，Japan）和扫描透射电镜（STEM，JEM-ARM200F，JEOL Co.，Japan）对样品的形貌和结构进行表征。通过 X 射线衍射（XRD，X'Pert，PAN analytical BV，the Netherlands）对晶体结构进行分析。通过 X 射线光电子能谱（XPS，PHI 5600，Perkin-Elmer Inc.，USA）对其化学成分进行表征。

TiO_2 SCs 的光化学性质通过此外可见漫反射光谱（DRS）、瞬态光电流响应进行表征，DRS 由紫外/可见分光光度计（UV 2550，Shimadzu Co.，Japan）进行测定。通过黑暗与紫外光照射下三电极体系中的 CV、差分脉冲阳极溶出伏安法（DPV）和 EIS 等来表征其电化学性质，通过傅里叶变换红外光谱（FTIR）和拉曼光谱等分析方法表征电极的抗污染性能。

3.2.2

双酚 A 与垃圾渗滤液的处理

在一个三电极单室电解池中进行 BPA 降解实验，反应温度约为 20 ℃（图 3.1）。

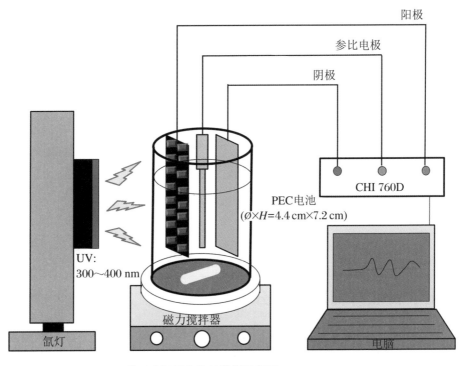

图 3.1　PEC-UV-TiO_2 体系降解污染物的装置示意图

阳极有效面积为 6.0 cm², TiO₂ SCs 负载量约为 0.05 mg·cm⁻²（共 0.30 mg），阴极采用面积相同的钛片。电极间距为 1.0 cm。采用电化学工作站（CHI 760D, Chenhua Co., China）对 80 mL 的 5~100 mg·L⁻¹ BPA 溶液（含 0.1 mol·L⁻¹ Na₂SO₄ 电解质）进行电解，施加的偏压（相对于 SCE）控制在 0.5~2.0 V 范围内。紫外光源为高压 500 W 氙灯（PLS-SXE500, Beijing Trust Tech Co., China）。UV 波长在 300~400 nm 范围内，光强约为 2.1 mW·cm⁻²，由光强计（Model FZ-A, Photoelectric Instrument Plant of Beijing Normal University, China）在距离氙灯中心 3 cm 处测量。在 PEC 体系中没有使用额外的外部电阻或红外补偿。

以 P25 为对照，在足够的剂量下进行 BPA 的标准光催化降解实验，催化剂常规用量为 1.0 g·L⁻¹ [2]。

下面开展垃圾渗滤液的降解处理实验。该垃圾渗滤液于 2016 年 5 月至 8 月期间从合肥市某市政垃圾填埋场收集，并保存在冰箱（4 ℃）中。原渗滤液采用好氧沉池法、反硝化法和活性污泥法进行了现场处理，去除可降解有机物和氨。经预处理的渗滤液的理化特性如表 3.1 所示。在紫外-可见光谱中，未发现 250~700 nm 处的特征峰。因此，选择 300 nm 作为颜色还原度的评价标准[24]。

表 3.1 PEC-UV-TiO₂ 体系所处理的垃圾渗滤液的指标参数

参数	范围	平均值
pH	7.0~7.5	7.3
电导率（mS·cm⁻¹）	13.7~20.0	16.4
化学氧（COD, mg·L⁻¹）	870~1210	1050
[NH₄]⁺（mg·L⁻¹）	130~310	220
凯氏法则氮化物总量（TKN, mg·L⁻¹）	210~450	340
硫化物（mg·L⁻¹）	430~520	480
氯化物（mg·L⁻¹）	95~187	140

3.2.3

电化学表征与辅助测试分析

紫外可见吸收光谱和荧光光谱分别通过紫外-可见分光光度计（UV-2401PC, Shimadzu Co., Japan）和荧光分光光度计（RF-5301PC, Shimadzu

Co.，Japan）进行采集。BPA 的测定通过高效液相色谱法（HPLC-1100，Agilent Co.，USA），通过 Hypersil-ODS 反相柱，VWD 检测器在 254 nm 处检测。流动相为水和甲醇的混合物（体积比为 30∶70），流速为 $1.0 \text{ mL} \cdot \text{min}^{-1}$。矿化效率由 TOC 分析仪（Vario TOC cube，Elementar Co.，Germany）来测定和计算。降解中间产物通过气相质谱（GCT Premier，Waters Inc.，USA）和液相质谱（LC-MS，LCMS-2010A，Shimadzu Co.，Japan）进行测定。通过标准方法测定 COD[24]。

通过 $0.1 \text{ mol} \cdot \text{L}^{-1}$ Na_2SO_4 溶液中的光电流测试和对苯二甲酸在紫外光照射下（200~400 nm）在三电极体系中·OH 数量的生成来评估光化学性质，以 TiO_2 SCs/碳纸作为工作电极，铂丝作为对电极，SCE 作为参比电极。光源为 500 W 氙灯（PLS-SXE500，Beijing Trusttech Co.，China）。CV 在 $0.1 \text{ mol} \cdot \text{L}^{-1}$ Na_2SO_4 和 5.0 mm $[Fe(CN)_6]^{3-}/[Fe(CN)_6]^{4-}$ 扫速为 $0.1 \text{ V} \cdot \text{s}^{-1}$ 的三电极电池中测定，DPV 在相同条件下的 $0.1 \text{ mol} \cdot \text{L}^{-1}$ Na_2SO_4 和 $30 \text{ mg} \cdot \text{L}^{-1}$ BPA 中测定。EIS 通过在 $0.1 \text{ mol} \cdot \text{L}^{-1}$ Na_2SO_4 水溶液中施加交流电压幅值为 5.0 mV、在 $10^5 \sim 10^{-2}$ Hz 的频率范围内、外加偏压和紫外光照来进行测量。

在粉末态 TiO_2 SCs、P25 和 $Ti_{0.7}Ru_{0.3}O_2$ 均匀负载的 GCE（直径为 5.0 mm、有效电极面积为 0.196 cm^2）制备过程中，使用了一种特定的有机导电黏结剂，该黏结剂能够改善 TiO_2 的分散性和机械强度，减小 GCE 与 TiO_2 之间的电荷转移阻力。将 2.0 mg TiO_2 SCs 分散在 2.0 mL 异丙醇溶剂中，超声分散 30 min，随后将 20 μL 分散液（5 μL，0.05 wt %）滴加到 GCE 电极表面，负载量约为 0.10 mg，在空气中放置，让溶剂自然挥发，直到催化剂均匀地分布在整个玻碳电极表面。

3.2.4
{001}晶面暴露率计算

制备的 TiO_2 和 TiO_{2-x} SCs 中高能{001}晶面的暴露率[42-46]为

$$S_{001} = 2a^2 \tag{3.1}$$

$$S_{101} = 8\left(\frac{1}{2}EG \times b - \frac{1}{2}EF \times a\right) \tag{3.2}$$

$$S_{001}(\%) = \frac{S_{001}}{S_{001} + S_{101}}$$

$$= \frac{2a^2}{2a^2 + 8\left(\frac{1}{2}EG \times b - \frac{1}{2}EF \times a\right)}$$

$$= \frac{a^2}{a^2 + 4\left(\frac{1}{2} \times \frac{\frac{1}{2}b}{\cos\theta} \times b - \frac{1}{2} \times \frac{\frac{1}{2}a}{\cos\theta} \times a\right)}$$

$$= \frac{a^2}{a^2 + \frac{b^2 - a^2}{\cos\theta}} = \frac{1}{1 + \frac{\frac{b^2}{a^2} - 1}{\cos\theta}}$$

$$= \frac{\cos\theta}{\cos\theta + \frac{b^2}{a^2} - 1} = \frac{\cos\theta}{\cos\theta + \left(\frac{a}{b}\right)^{-2} - 1} \tag{3.3}$$

锐钛矿 TiO_2 SCs 中的 θ 为{001}和{101}之间夹角的角度理论值(68.3°)，在它的 SLAB 模型中，两个独立的参数 a 和 b 分别表示正方形{001}"截断"晶面的边长和双锥体的边长。

3.3
紫外光辅助 TiO_2 单晶电催化降解双酚 A 的效能与机理分析

3.3.1
光电耦合催化性能评估

TiO_2 SCs 的形态特征和结构如图 3.2 所示，制备的双锥体 TiO_2 SCs 呈片状，由 8 个等效{101}晶面和 2 个等效{001}晶面包裹（图 3.2(a、b)），清晰的晶格条纹和折射率{001}带轴衍射进一步证实了其单晶结构（图 3.2(a~e)）[28-29]。X 射线衍射结果与锐钛矿 TiO_2（JCPDS No. 21-1272）高度匹配，衍射峰增宽主要是由于其晶体尺寸较小。通过分散滴涂法将 TiO_2 SCs 均匀负载在碳纸电极上,负载密度为 0.05 mg·cm^{-2},从图 3.2(f)中看出,碳纸基底与 TiO$_2$ SCs 通过物理吸附、静电键合或电荷转移发生相互作用。

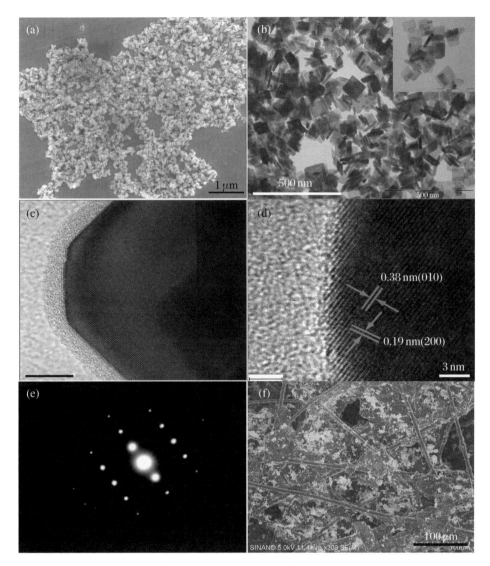

图 3.2 TiO$_2$ SCs 的 SEM(a)、TEM(b)、HRTEM(c、d)和 SAED(e)及反应后电极的 SEM(f)表征

一种高效的 PEC 阳极材料应该同时具备优异的 PC 活性和 EC 活性。TiO$_2$ SCs 优异的 PC 活性(图 3.3(a、b))、EC 活性(图 3.3(c、d))和 PEC(图 3.3(e、f))活性在光谱吸收、电子转移表征、光电流响应以及 PL 测试·OH 捕获和与 P25、Ti$_{0.7}$Ru$_{0.3}$O$_2$ 电极的污染物降解(图 3.9 和图 3.10)性能对比等实验中得到了充分体现。由 CV 测试中 ΔE_p 的减小,可以看出 TiO$_2$ SCs 发生电子转移的动力学势垒降低、界面反应活性面积提高(图 3.3(c)),这两者对提高晶面调控工程和单晶结构催化剂的 EC 活性至关重要[35]。TiO$_2$ SCs 电极对 BPA 阳极氧化的峰电

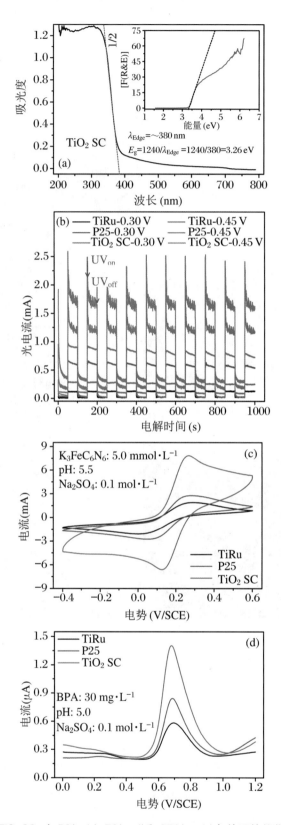

图 3.3 TiO$_2$ SCs 在 PC(a、b)、EC(c、d) 和 PEC(e~h) 条件下的催化活性表征

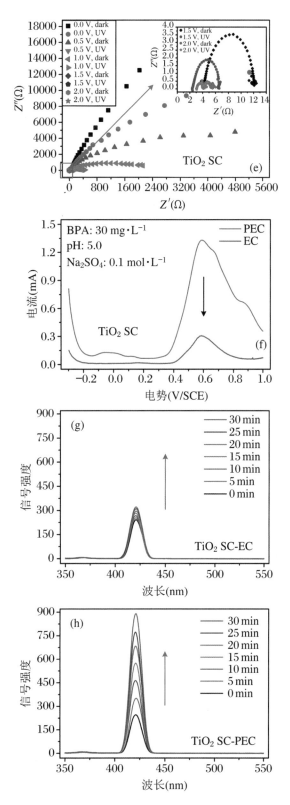

续图 3.3 TiO$_2$ SCs 在 PC(a、b)、EC(c、d) 和 PEC(e~h) 条件下的催化活性表征

流值较大,进一步证实了其具有较高的 EC 活性(图 3.3(d))[36]。这些电化学优势主要归功于单晶结构和暴露的高能{001}晶面[24-27]。

此外,PC 与 EC 耦合后,PEC 反应的电子转移阻力进一步降低,BPA 的转化率和·OH 产率均显著提高(图 3.3(e~h))。这些特性使{001}-TiO_2 SCs 成为高效降解污染物的优异 PEC 材料。

3.3.2
双酚 A 降解效能评估

BPA 在 PEC($k=26.92\times10^{-3}$ min^{-1})体系中的降解速率最大,远远高于在 PC($k=0.41\times10^{-3}$ min^{-1})和 EC($k=19.10\times10^{-3}$ min^{-1})中的降解速率之和(图 3.4(a))。在 PEC 中,30 $mg·L^{-1}$ 的 BPA 能在 2.0 h 被完全降解,而在 EC 体系中达到相同降解效果需要 4.0 h。当 EC 和 PC 在 TiO_2 SCs 电极上结合时,PEC 体系中 BPA 降解效率显著提高,说明可能存在有效的协同作用。

虽然少量的水分子可能会在 +1.3 V/SCE 电压下发生氧化,但由于过电势的缘故,这种反应可以忽略不计,从 LSV 的扫描结果和 TiO_2 SCs 电极在 PEC 反应过程中的直观照片来看(图 3.5),在溶液表面并没有明显的气泡(氧气)生成。

需要指出的是,PC 中观察到的较低的 BPA 去除率并不表明电极的光驱动活性较弱,而是催化剂负载量不足所致。在此 PC 体系中光催化剂的溶液质量/体积为 0.00375 $mg·mL^{-1}$,仅为 PC 相关研究典型值(1.0 $mg·mL^{-1}$)的 1/300 左右[2]。

与相同条件下的 P25 和 $Ti_{0.7}Ru_{0.3}O_2$ 电极相比,TiO_2 SCs 电极在 BPA 降解和矿化方面同时具有较高的 EC 和 PEC 活性(图 3.4(b~f))。由于单位时间内的 BPA 去除量存在上限,降解速率常数随着底物浓度的增加而不断降低。以上结果表明,虽然 TiO_2 是一种典型的半导体光催化剂,但当其晶体形貌和暴露晶面经过精细的调控后,同样可以成为一种良好的电催化剂(图 3.2)。TiO_2 SCs 的电催化优势主要在于其单晶结构和高暴露的{001}晶面[24]。通过使单个 TiO_2 SCs 电极上同时具有较高的 EC 和 PC 活性,有效地建立了一种新的以 EC 主导的 PEC 体系,与其他报道中的以 PC 主导的 PEC 体系相比,它能够在较低的偏压下更为高效地降解污染物(图 3.4)[32-34]。此外,在 EC 和 PEC 体系中,双功能 TiO_2 SCs 电极的活性也远远高于碳纸和 Sb-SnO_2/Ti 电极。

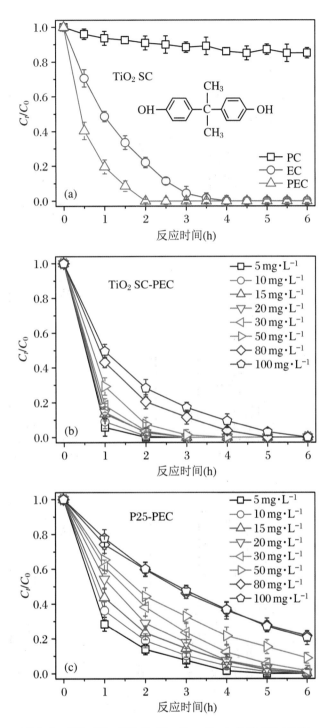

图 3.4 TiO_2 SCs、P25 和 $Ti_{0.7}Ru_{0.3}O_2$ 电极在 PC、EC 和 PEC 三种体系中对 BPA 的催化降解性能对比:降解(a~e) 和矿化(f)

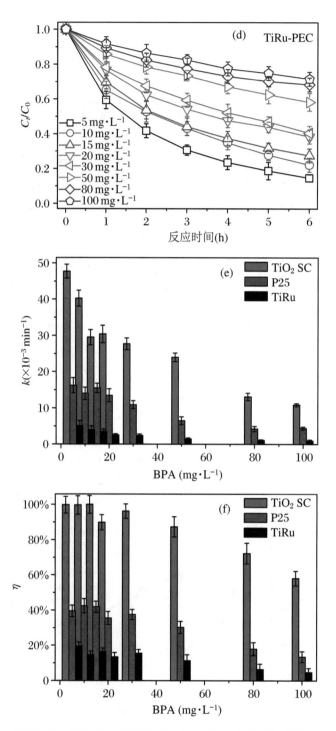

续图 3.4 　TiO$_2$ SCs、P25 和 Ti$_{0.7}$Ru$_{0.3}$O$_2$ 电极在 PC、EC 和 PEC 三种体系中对 BPA 的催化降解性能对比：降解(a～e)和矿化(f)

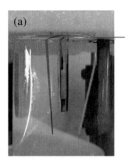 

图 3.5 TiO$_2$ SCs 电极在 PEC 条件下降解 BPA 过程中的阳极照片

在 BPA 降解过程中,我们检测到了 4-异丙基苯酚(图 3.6(a))、对苯二酚(图 3.6(b))、苯醌(图 3.6(c))、柠檬酸(图 3.6(d))、马来酸(图 3.6(e))和乙酸(图 3.6(f))等多种中间产物。在 PEC 体系中对苯二酚和顺丁烯二酸的最大积累浓度较高,达到峰值的时间较 EC 短(图 3.6(b、e)),说明 PEC 体系中 BPA 的降解速率较快。此外,其他四种中间产物的最大富集浓度在 EC 中均高于 PEC,且达到峰值的时间较 PEC 长(图 3.6(a、c、d、f)),表明 EC 中 BPA 的矿化速率较慢。这些结果与 TiO$_2$ SCs 电极在 PEC 体系中降解污染物的良好活性高度一致(图 3.4(a))。

以上结果还表明,PEC 体系的 BPA 去除率和电流效率明显高于 EC(表 3.2)。需要指出的是,PC 体系的能耗效率未能提供,因为 TOC 去除率过低,无法准确测定其数值。由于使用的氙灯能量输入高,紫外输出低,因此 PEC 的能耗仅通过组合电化学过程计算。在这种情况下,计算出的能量消耗只占 PEC 总能量消耗的一小部分,这是因为没有考虑紫外光源照射的能量消耗(表 3.2)。此外,电辅助吸附也有助于去除 BPA,通过在 TiO$_2$ SCs 电极上建立协同吸附促进降解[37]。从 20 次循环降解实验的结果可以看出,PEC 体系具有比 EC 好得多的 BPA 降解矿化的优异循环稳定性(图 3.7(a、c、d))。在 EC 体系中,循环到第 7 次的时候,催化活性就开始显著下降,这也说明在低偏压条件下 TiO$_2$ 的电极表面逐步受到污染,中间体聚合物累积得越来越多(图 3.7(b))。

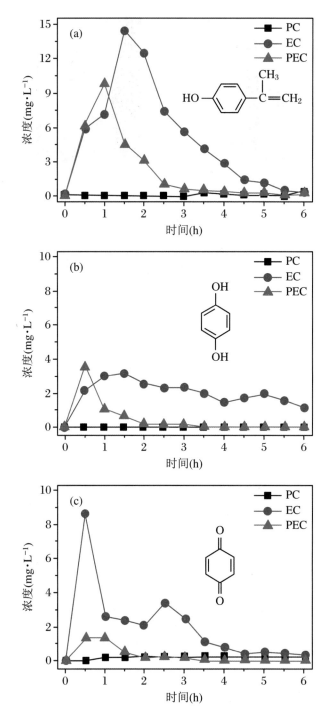

图 3.6 TiO₂ SCs 电极在 PC、EC 和 PEC 条件下降解 BPA 的中间产物分析：4-异丙基苯酚(a)、对苯二酚(b)、苯醌(c)、柠檬酸(d)、马来酸(e)和乙酸(f)

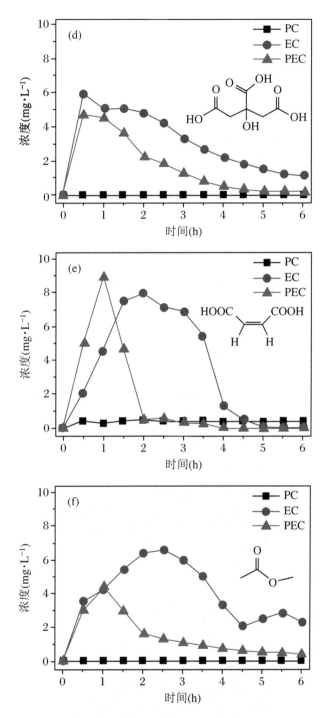

续图 3.6　TiO$_2$ SCs 电极在 PC、EC 和 PEC 条件下降解 BPA 的中间产物分析：4-异丙基苯酚(a)、对苯二酚(b)、苯醌(c)、柠檬酸(d)、马来酸(e)和乙酸(f)

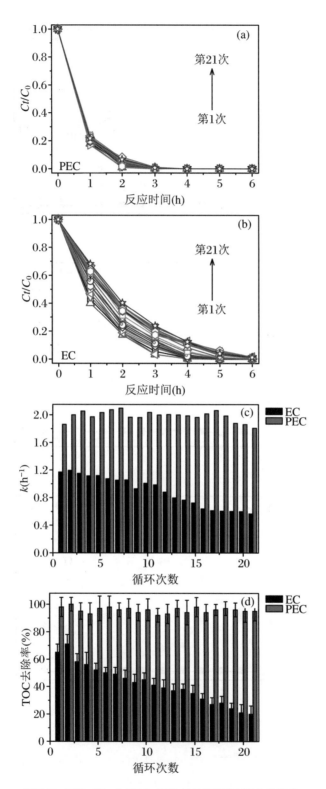

图 3.7 TiO$_2$ SCs 在 EC 和 PEC 条件下降解 BPA 的稳定性对比：降解(a、b)、动力学(c)和矿化(d)

表 3.2 TiO₂ SCs 电极在 PC、EC 和 PEC 条件下对 BPA 的降解实验参数①

过程	电势(V, vs SCE)	浓度 (mg·L^{-1})	去除率②	动力学常数 (k_s, ×10^{-3} min^{-1})③	矿化效率④	MCE⑤	AE (Wh·mg TOC^{-1})⑥
PC	—	5	12.40%	0.77	—⑦	—	—
	—	10	10.96%	0.61	—⑦	—	—
	—	15	10.05%	0.60	—⑦	—	—
	—	20	8.17%	0.49	—⑦	—	—
	—	30	7.33%	0.41	—⑦	—	—
	—	50	3.32%	0.24	—⑦	—	—
	—	80	2.25%	0.21	—⑦	—	—
	—	100	1.04%	0.09	—⑦	—	—
EC	0.5	30	16.77%	0.54	14.23%	53.23%	0.021%
	0.8	30	13.06%	0.90	12.82%	46.60%	0.038%
	1.0	30	20.74%	6.29	18.14%	23.06%	0.096%
	1.3	5	98.72%	35.57	66.57%	3.98%	2.867%
	1.3	10	87.45%	21.98	61.02%	7.82%	1.459%
	1.3	15	69.33%	16.34	48.47%	9.36%	1.219%
	1.3	20	62.87%	19.48	40.50%	10.67%	1.069%
	1.3	30	77.38%	19.10	66.76%	20.86%	0.547%
	1.3	50	72.64%	16.23	48.39%	32.51%	0.351%
	1.3	80	40.07%	6.65	24.34%	27.32%	0.418%
	1.3	100	41.95%	5.17	18.87%	25.57%	0.446%
	1.5	30	97.89%	29.51	74.52%	16.58%	0.794%
	1.8	30	100.00%	36.86	86.07%	11.39%	1.387%
PEC	0.5	30	26.40%	0.70	24.65%	70.85%	0.016%
	0.8	30	29.79%	2.00	28.92%	71.36%	0.024%
	1.0	30	42.28%	12.54	34.71%	44.12%	0.050%
	1.3	5	100.00%	47.75	99.83%	5.72%	1.987%
	1.3	10	100.00%	40.24	99.67%	11.85%	0.959%
	1.3	15	100.00%	29.50	100.00%	17.43%	0.652%
	1.3	20	98.25%	30.41	89.87%	23.40%	0.485%
	1.3	30	98.81%	26.92	96.20%	31.37%	0.362%
	1.3	50	97.98%	23.99	87.36%	52.52%	0.216%
	1.3	80	80.83%	13.09	72.32%	61.08%	0.186%
	1.3	100	72.17%	10.87	58.04%	65.70%	0.173%
	1.5	30	100.00%	26.62	99.26%	24.79%	0.529%
	1.8	30	100.00%	39.25	99.83%	13.96%	1.126%

注:① 反应平行进行三次,计算出平均值,反应条件:溶液体积:80 mL,BPA 浓度:5～100 mg·L^{-1},Na₂SO₄ 浓度:0.1 mol·L^{-1},阳极尺寸:6.0 cm²,催化剂负载量:0.05

mg·cm^{-2}×6.0 cm^2=0.30 mg,阴极尺寸:6.0 cm^2,钛片,电极间距:1.0 cm,外加偏压:+0.5~+1.8 V/SCE,紫外光源:500 W 氙灯,λ<420 nm,不调控 pH,温度:~20 ℃,反应时间:2.0/6.0 h。

② 反应时间 2.0 h。

③ $\ln\left(\dfrac{C_0}{C_t}\right) = k_s \times t$。

④ 反应时间:6.0 h。

⑤ MCE(矿化电流效率)=$\dfrac{\Delta \text{TOC}_{\text{exp}}}{\Delta \text{TOC}_{\text{theor}}} \times 100\%$。

⑥ AE = $\dfrac{U_{\text{cell}} \times I \times \Delta t}{(\text{TOC}_0 - \text{TOC}_t) \times V}$,电化学过程的计算中,只包含了 PEC 体系中的一小部分能耗,因为紫外光辐照的能耗未算在内。

⑦ 因数值太低而难以准确测定。

3.3.3 电极防污策略

从 DPV 谱图中可以看出,在 EC 的第一个循环后,BPA 的氧化峰(0.60 V)大幅降低,甚至在此后的循环中几乎完全消失(图 3.8(a)),这说明阳极聚合物已开始逐渐积累,使 TiO$_2$ 电极迅速失去催化活性[1,31]。相比之下,由于 PEC 具有更强的催化能力和防污性能,这种失活现象在 PEC 中大幅减少(图 3.8(b))。

用红外光谱(FTIR)和紫外可见红外吸收光谱(DRS)对反应后电极样品的甲醇浸取液进行了表征。与 BPA 标准品相比,在 EC 反应电极浸取液的特征光谱中(图 3.8(c)),O—H 键在羟基区域变宽,C═C 环、平面外 C—H 键发生弯曲,在 3220 cm^{-1}、1600 cm^{-1} 和 830 cm^{-1} 处,直至消失,但在 2924 cm^{-1} 和 2854 cm^{-1} 处存在两个较强的尖峰,符合 sp3 杂化的 C—H 键拉伸模式的特点,而非共轭 C═O 键在 1750 cm^{-1} 处可以观察到[31]。在 DRS 光谱中,酚类在 270 nm 处关联 π-π 电子传递的 B 带消失,只观察到 227 nm 处的 E_2 带(图 3.8(d))。这些结果表明,EC 电极上的聚合物主要由包括羰基在内的脂肪族碳氢化合物组成[31]。相比之下,从 PEC 反应电极浸取液中并没有检测到 FTIR 和 DRS 的明显特征光谱,这意味着 TiO$_2$ SCs 电极上的中间体聚合物几乎被全部降解,活性位点完全再生,从而获得了较高的电化学活性和稳定性。

通过检测作为氧化还原标记物的铁氰化钾溶液(10 mmol·L^{-1})的 CV 电化学响应,来评估 TiO$_2$ SCs 电极的电化学稳定性(图 3.8(e))。使用过的 EC 电

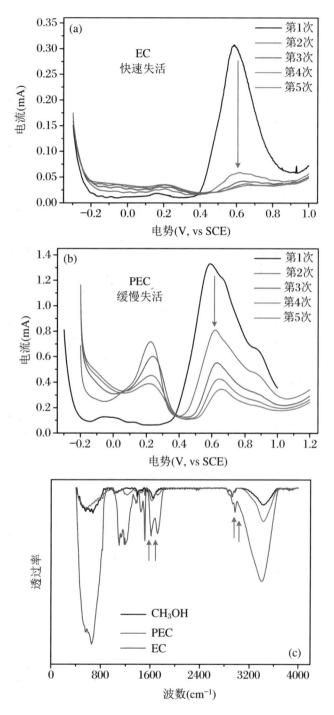

图3.8　TiO_2 SCs 电极在 EC 和 PEC 体系中抗污性能的对比：DPV(a、b)、FTIR(c)、DRS(d)、CV(e)和 Raman(f)

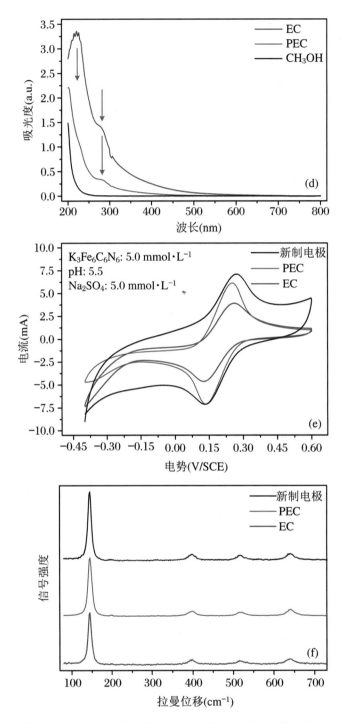

续图 3.8 TiO₂ SCs 电极在 EC 和 PEC 体系中抗污性能的对比：DPV(a、b)、FTIR(c)、DRS(d)、CV(e) 和 Raman(f)

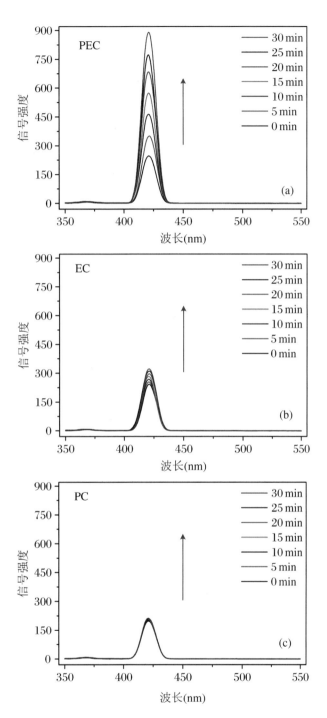

图 3.9　TiO$_2$ SCs、P25 和 Ti$_{0.7}$Ru$_{0.3}$O$_2$ 电极在 PEC(a、d、g)、EC(b、e、h) 和 PC(c、f、i) 三种体系中的·OH 捕获荧光测试

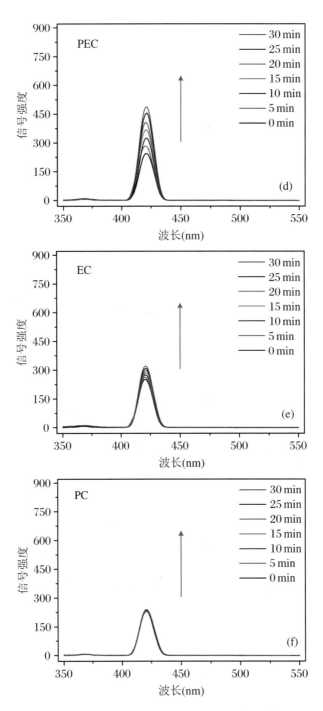

续图 3.9 TiO$_2$ SCs、P25 和 Ti$_{0.7}$Ru$_{0.3}$O$_2$ 电极在 PEC (a、d、g)、EC(b、e、h)和 PC(c、f、i)三种体系中的·OH 捕获荧光测试

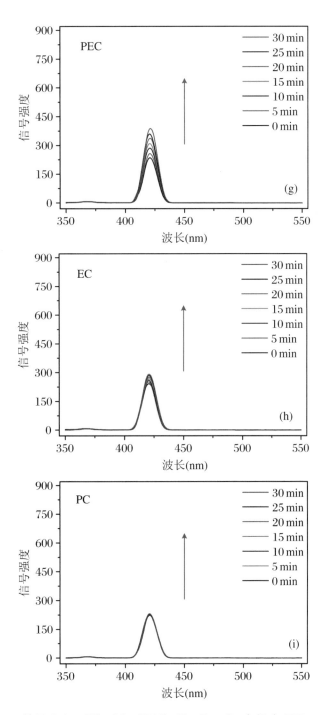

续图3.9 TiO$_2$ SCs、P25 和 Ti$_{0.7}$Ru$_{0.3}$O$_2$ 电极在 PEC(a、d、g)、EC(b、e、h)和 PC(c、f、i)三种体系中的·OH 捕获荧光测试

极(3.92 mA,扫速为 10 mV·s^{-1})上的氧化电流明显低于原始电极(7.11 mA)上的氧化电流,表明由于 EC 活性有限,中间体聚合物的非彻底矿化导致阳极失活。相比之下,PEC(降解 BPA 后的)电极的氧化电流下降幅度要小得多(6.14 mA),这意味着由于 PEC 活性足够高,没有大量有机物残留。这些结果与 FTIR 和 DRS 的结果完全吻合。此外,我们还利用拉曼分析探讨了循环降解实验后催化剂的晶体结构和电极稳定性变化。在 142 cm^{-1}、394 cm^{-1}、512 cm^{-1} 和 634 cm^{-1} 左右可以看到拉曼特征峰(图 3.8(f)),EC 和 PEC(反应后的电极)经过 5 次循环降解后均未发生增宽和移位,由此说明 TiO$_2$ SCs 并未发生结构破坏。

因此,TiO$_2$ SCs 电极在 EC 中的失活主要是由 1.3 V 时电化学驱动力有限、吸附在活性和非活性位点上的中间体聚合物所致,并非阳极材料的晶体结构遭到破坏。DRS 和 FTIR 的测试结果进一步证实了这一点。相比之下,由于 EC 和 PC 对双功能 TiO$_2$ SCs 阳极的协同作用,PEC 过程未发生电极失活和结构破坏。

此外,我们还开展了粉末态标准光催化降解,实验结果如图 3.10 所示。在相同条件下,高暴露{001}晶面 TiO$_2$ SCs 的紫外光催化降解活性和矿化能力均明显高于对照材料 P25。

TiO$_2$ SCs 低偏压电催化降解污染物的电极防污能力主要得益于其优异的 PC 性能(图 3.8)。同时,TiO$_2$ SCs 电极具备有效结合 PC 与 EC 的双功能性质,形成以独特的 EC 主导的 PEC 催化机理。虽然在典型 1.0 g·L^{-1} 的催化剂用量下,TiO$_2$ SCs 的 PC 活性比 P25 高(图 3.10),但在 BPA 降解和矿化方面仍低于 PEC 体系(图 3.4(a))。这一结果表明 PEC 体系中的 PC 与 EC 存在协同作用,其中发生了独特的 EC 主导的 PEC 机理,实现了污染物的高效降解(图 3.7、图 3.8 和图 3.13)。

3.3.4
反应机理及双酚 A 降解路径分析

双功能 TiO$_2$ SCs 可以分别通过非带隙紫外光源和带隙激发机理同时活化(图 3.12)。在 +1.3 V/SCE 的偏压下($E \leqslant \sim 3.2$ eV),电子从导带(Conduction Band,CB)激发到阴极残留空穴 CB(反应(3.4)),同时在紫外光($\lambda = 300 \sim 400$ nm)照射下电子从价带(Valance Band,VB)激发到 CB 形成活性电子-空穴对(反应(3.5))。然后,载流子因阳极偏压而形成空间上的分离,这在电极内部提供了一

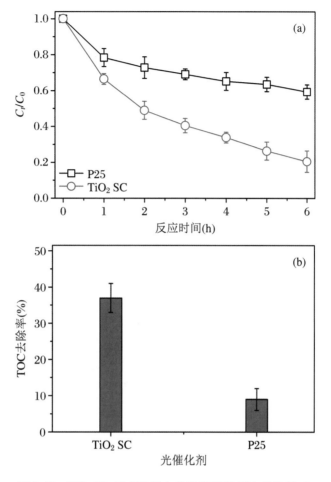

图 3.10 TiO₂ SCs 和 P25 粉末光催化降解 BPA 的性能对比：降解(a)和矿化(b)

个电位梯度,使电子沿着导电基底(反应(3.6)和反应(3.7))流向电极[38-39]。在水氧化作用下(反应(3.8)～反应(3.11)),TiO₂ SCs 上的电生空穴和光生空穴分别能形成表面吸附态和游离态的·OH[2]。电位偏压下的电生空穴和表面吸附态·OH 可以氧化污染物(≡C—OH)但不彻底矿化,从而生成各种中间体聚合物(反应(3.12)和反应(3.13)),这些产物可以被光生空穴和在紫外光 UV 照射下产生的自由·OH 有效降解矿化(反应(3.14)和反应(3.15))。当偏压进一步增大时,会发生以表面吸附态和游离态·OH 为主要中间产物的水氧化过程,因而导致了析氧反应的发生(反应(3.16)和反应(3.17))[1]：

$$TiO_2 + 电势 \rightarrow TiO_2(h^+)_{CB} + e^- （非带隙激发） \quad (3.4)$$

$$TiO_2 + UV \rightarrow TiO_2(h^+)_{VB} + TiO_2(e^-)_{CB} （带隙激发） \quad (3.5)$$

$$TiO_2(h^+)_{CB} + e^- + 电势 \rightarrow TiO_2(h^+)_{CB} + 电流 \quad (3.6)$$

$$TiO_2(h^+)_{VB} + TiO_2(e^-)_{CB} + 电势 \rightarrow TiO_2(h^+)_{VB} + 电流 \quad (3.7)$$

$$TiO_2(h^+)_{CB} + OH^- \rightarrow TiO_2(\cdot OH)_{bound} \quad (3.8)$$

$$TiO_2(h^+)_{CB} + H_2O \rightarrow TiO_2(\cdot OH)_{bound} + H^+ + e^- \quad (3.9)$$

$$TiO_2(h^+)_{VB} + OH^- \rightarrow TiO_2(\cdot OH)_{free} \quad (3.10)$$

$$TiO_2(h^+)_{VB} + H_2O \rightarrow TiO_2(\cdot OH)_{free} + H^+ + e^- \quad (3.11)$$

$$TiO_2(h^+)_{CB} + \equiv C-OH \rightarrow \equiv C-O\cdot \rightarrow \equiv C-O\cdot\cdot O-C \equiv \Rightarrow 聚合物 \quad (3.12)$$

$$TiO_2(\cdot OH)_{bound} + \equiv C-OH \rightarrow \equiv C-O\cdot \rightarrow \equiv C-O\cdot\cdot O-C \equiv \Rightarrow 聚合物 \quad (3.13)$$

$$TiO_2(h^+)_{VB} + 聚合物 \rightarrow 开环中间体 \rightarrow CO_2 + H_2O \quad (3.14)$$

$$TiO_2(\cdot OH)_{free} + 聚合物 \rightarrow 开环中间体 \rightarrow CO_2 + H_2O \quad (3.15)$$

$$TiO_2(\cdot OH)_{bound} \rightarrow 1/2 O_2 \uparrow + H^+ + e^- \quad (3.16)$$

$$TiO_2(\cdot OH)_{free} \rightarrow 1/2 O_2 \uparrow + H^+ + e^- \quad (3.17)$$

在所构建的体系中，BPA 的 PEC 降解主要是由 TiO_2 SCs 阳极表面的 EC 和 PC 共同引起的（图 3.3）。在外加偏压下 BPA 主要通过直接氧化而发生降解，这是阳极表面电生空穴的直接电子转移所致（路径 1）。既定的阳极电位（+1.3 V/SCE）明显高于 E^0（BPA/·BPA$^-$）（~+0.7 V/SCE、图 3.3（d,f）），但明显低于 E^0（H_2O/·OH）（~+2.3 V/SCE）[1,31]。EC 体系中的高催化活性（BPA 降解中的 k_{EC}/k_{PEC} 比值）和低 PL 信号（·OH 生成速率 k_{EC}/k_{PEC} 比值）表明这是一个 EC 主导的 PEC 协同机理与电生空穴介导的 BPA 降解路径，而其他三个氧化途径只扮演次要角色（图 3.4），这种在特异性调控晶面 TiO_2 SCs 电极上的 EC 主导的 PEC 机理与广泛报道的其他基于 TiO_2 的电极上的 PC 主导的 PEC 机理有根本差别[32-34]。在 +1.3 V/SCE 时，虽然电生空穴不能氧化吸附的 H_2O 和 OH^- 生成游离的 ·OH，但在本条件中不能排除电化学表面结合 ·OH 介导的间接氧化途径，因为这种电位偏压足以使表面结合 ·OH 氧化水分子（图 3.9）[1]。在电化学直接氧化（路径 1 和图 3.11）过程中，吸附的 BPA 通过电生空穴初步转化为 4-异丙基苯酚（$m/z = 133$）和苯酚。然后，以最直接、最简单的途径，通过芳香环的清除，将两种芳香中间产物部分氧化为脂肪酸。最终形成的脂肪酸完全矿化为 CO_2 和 H_2O[24]。但众所周知，芳香族化合物在电化学直接氧化过程中大量生成并逐渐积累了一些聚合产物[1]，这些产物通过直接阳极机理具有很强的抗氧化能力[31]。此外，这些聚合产物可以被强吸附到阳极表面，并牢牢地占据活性位点阻碍进一步反应的发生，导致电极严重失活（图 3.8（a）和图 3.11）[1,31]。因此，在 EC 中只有部分不充分的 BPA 的氧化过程发生

(图 3.4、图 3.6 和图 3.7)。

由于光生空穴在带隙激发下具有足够高的电位(E_{VB} = ~2.8 eV/SHE)[2],BPA 在紫外光照射下通过光生空穴引发的光化学直接氧化(途径 1 和图 3.11)和光生空穴引发的光化学间接氧化(途径 2 和图 3.11)而降解。·OH 介导的氧化有三个代表性步骤(途径 2 和图 3.11):首先,BPA 的一些羟基化衍生物由于游离·OH 对 BPA 分子的不同碳原子的选择性攻击而迅速生成,而这些芳香衍生物不稳定,易通过异丙基桥式裂解分解为单环芳香族化合物;随后,这些芳香族化合物进一步发生环裂解,形成短链脂肪酸;最后,这些有机酸被进一步氧化为 CO_2 和 H_2O,实现 BPA 的彻底矿化[40]。

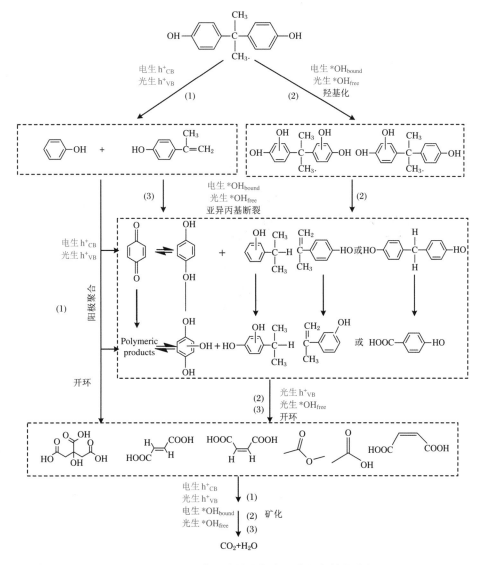

图 3.11　BPA 在 PEC-UV-TiO₂ SCs 体系中的降解中间产物与转化路径图

在路径 1 中，BPA（通过电子传递）直接氧化产生的中间产物的积累使阳极发生污染，从而导致 EC 性能不断降低（图 3.6 和图 3.7）[1,31]。当转化率有限的 EC 与 PC 结合在双功能 TiO$_2$ SCs 上构建为 PEC 协同体系时，很容易发生完全、稳定和非选择性的 BPA 降解（图 3.6 和图 3.7）。主要是因为这些在电化学直接氧化（路径 1）中形成的高抗性、强吸附性的聚合产物也可以被氧化为脂族酸，最终通过游离态·OH 介导的光化学间接氧化矿化为 CO$_2$（路径 3 和图 3.11）[3-7]。

尽管单独的 PC 只表现出微弱的活性（图 3.4(a)），但其在 EC 的辅助下产生的 PC-EC 协同作用可能有利于提升 PEC 体系的光化学防污能力（图 3.9）[32-34]。因此，TiO$_2$ SCs 表面被占用活性位点迅速释放，EC 阳极材料因 PC 形成的·OH$_{free}$ 而迅速再生（图 3.13(b)）。反过来，这两个过程会加速污染物降解，因而使 PEC 成为一个优异而稳定的 BPA 降解体系（表 3.2 和图 3.6）。

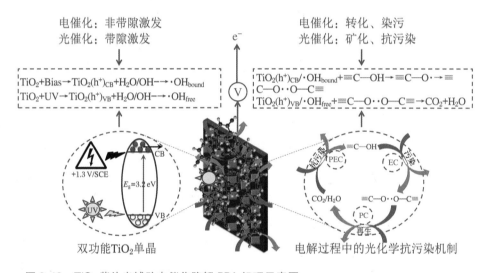

图 3.12　TiO$_2$ 紫外光辅助电催化降解 BPA 机理示意图

3.3.5

垃圾渗滤液处理效果

为了验证 PEC 实际应用的可行性，尝试用该体系处理垃圾渗滤液（含有多种难降解污染物的有毒废水）[41]，经过 7.0 h 降解后，虽然 PEC/EC 的脱色效果差异不大（图 3.13(a)），但 PEC 的 COD 去除率（＞75%）明显优于 EC（＜40%）（图 3.13(b)），说明 TiO$_2$ SCs/PEC 体系可以作为一种处理实际复杂废水的有效方法。

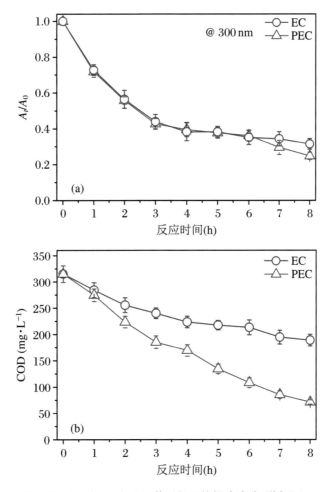

图 3.13　PEC/EC-TiO$_2$ SCs 体系处理垃圾渗滤液：脱色(a)、COD 去除(b)

本章小结

在本章工作中，借助富含{001}高能暴露晶面的 TiO$_2$ 单晶同时具有的优异光催化和电催化活性的特点，构建紫外光耦合电催化氧化新体系进行污染物的协同降解，并发展了阳极抗污染的新策略。利用光辅助电催化体系中·OH$_{free}$ 所介导的优异光化学氧化能力，在低偏压条件下实现了有机污染物的高效降解和垃圾渗滤液的有效处理。拉曼光谱、红外光谱、吸收光谱和电化学表征实验结果表明,在污染物降解过程中TiO$_2$电极表面始终保持相对洁净,说明污染物低偏压阳极氧化产生的有机聚合物可以通过游离态、高活性·OH介导的光化学氧化途径得到有效去除。基于单晶电极优异紫外光催化活性,本章提出了光化学阳极抗污染思路,有效解决了污染物低压电化学转化过程中的电极污染问题,并构建了以电催化为主导、光催化为辅助的光电耦合催化体系,显著提升了低偏压条件下TiO$_2$对污染物的电催化矿化效率。该工作为研发实用的高效水处理技术开辟了新的途径。

参考文献

[1] Panizza M, Cerisola G. Direct and mediated anodic oxidation of organic pollutants[J]. Chem. Rev., 2009, 109: 6541-6569.

[2] Chong M N, Jin B, Chow C W K, et al. Recent developments in photocatalytic water treatment technology: a review[J]. Wat. Res., 2010, 44: 2997-3027.

[3] Hu L S, Fong C C, Zhang X M, et al. Au nanoparticles decorated TiO$_2$ nanotube arrays as a recyclable sensor for photo-enhanced electrochemical detection of bisphenol A[J]. Environ. Sci. Technol., 2016, 50: 4430-4438.

[4] Zhang C L, Xu J Q, Li Y T, et al. Photocatalysis-induced renewable field-effect transistor for protein detection[J]. Anal. Chem., 2016, 88: 4048-4054.

[5] Xu J Q, Liu Y L, Wang Q, et al. Photocatalytically renewable micro-electrochemical sensor for real-time monitoring of cells[J]. Angew. Chem., Int. Ed., 2015, 54: 14402-14406.

[6] Xu J Q, Duo H H, Zhang Y G, et al. Photochemical synthesis of shape-controlled nanostructured gold on zinc oxide nanorods as photocatalytically renewable sensors[J]. Anal. Chem., 2016, 88: 3789-3795.

[7] Hu L S, Huo K F, Chen R S, et al. Recyclable and high-sensitivity electrochemical biosensing platform composed of carbon-doped TiO$_2$ nanotube arrays[J]. Anal. Chem., 2011, 83: 8138-8144.

[8] Pelegrini R T, Freire R S, Duran N, et al. Photoassisted electrochemical degradation of organic pollutants on a DSA type oxide electrode: process test for a phenol synthetic solution and its application for the E1 bleach kraft mill

effluent[J]. Environ. Sci. Technol., 2001, 35: 2849-2853.

[9] Malpass G R P, Miwa D W, Miwa A C P, et al. Photo-assisted electrochemical oxidation of atrazine on a commercial Ti/Ru$_{0.3}$Ti$_{0.7}$O$_2$ DSA electrode[J]. Environ. Sci. Technol., 2007, 41: 7120-7125.

[10] Qu J H, Zhao X. Design of BDD-TiO$_2$ Hybrid electrode with P-N function for photoelectroatalytic degradation of organic contaminants[J]. Environ. Sci. Technol., 2008, 42: 4934-4939.

[11] Asmussen R M, Tian M, Chen A C. A new approach to wastewater remediation based on bifunctional electrodes[J]. Environ. Sci. Technol., 2009, 43: 5100-5105.

[12] Chai S N, Zhao G H, Zhang Y N, et al. Selective Photoelectrocatalytic degradation of recalcitrant contaminant driven by an n-P heterojunction nanoelectrode with molecular recognition ability [J]. Environ. Sci. Technol., 2012, 46: 10182-10190.

[13] Li G T, Qu J H, Zhang X W, et al. Electrochemically assisted photocatalytic degradation of Acid Orange 7 with β-PbO$_2$ electrodes modified by TiO$_2$[J]. Wat. Res., 2006, 40: 213-220.

[14] Wang P F, Cao M H, Ao Y H, et al. Investigation on Ce-doped TiO$_2$-coated BDD composite electrode with high photoelectocatalytic activity under visible light irradiation[J]. Electrochem. Comm., 2011, 13: 1423-1426.

[15] Li P Q, Zhao G H, Li M F, et al. Design and high efficient photoelectric-synergistic catalytic oxidation activity of 2D macroporous SnO$_2$/1D TiO$_2$ nanotubes[J]. Appl. Catal., B, 2012, 111-112: 578-585.

[16] Quan X, Chen S, Su J, et al. Synergetic degradation of 2, 4-D by integrated photo-and electrochemical catalysis on a Pt doped TiO$_2$/Ti electrode[J]. Sep. Purif. Technol., 2004, 34: 73-79.

[17] Li Y H, Liu P F, Pan L F, et al. Local atomic structure modulations activate metal oxide as electrocatalyst for hydrogen evolution in acidic water [J]. Nat. Commun., 2015, 6: 8064.

[18] Zhang J, Xu Q, Feng Z C, et al. Importance of the relationship between surface phases and photocatalytic activity of TiO$_2$[J]. Angew. Chem., Int. Ed., 2008, 47: 1766-1769.

[19] Zhao X, Qu J H, Liu H J, et al. Photoelectrocatalytic degradation of triazine-containing azo dyes at γ-Bi$_2$MoO$_6$ film electrode under visible light

irradiation($\lambda > 420$ nm)[J]. Environ. Sci. Technol., 2007, 41: 6802-6807.

[20] Zhao X, Zhu Y F. Synergetic Degradation of rhodamine B at a porous $ZnWO_4$ film electrode by combined electro-oxidation and photocatalysis[J]. Environ. Sci. Technol., 2006, 40: 3367-3372.

[21] Zhao X, Xu T G, Yao W Q, et al. Photoelectrocatalytic degradation of 4-chlorophenol at Bi_2WO_6 nanoflake film electrode under visible light irradiation[J]. Appl. Catal., B, 2007, 72: 92-97.

[22] Chen X B, Mao S S. Titanium dioxide nanomaterials: synthesis, properties, modifications, and applications[J]. Chem. Rev., 2007, 107: 2891-2959.

[23] Liu L, Chen X B. Titanium dioxide nanomaterials: self-structural modifications[J]. Chem. Rev., 2014, 114: 9890-9918.

[24] Zhang A Y, Long L L, Liu C, et al. Electrochemical degradation of refractory pollutants using TiO_2 single crystals exposed by high-energy {001} facets[J]. Wat. Res., 2014, 66: 273-282.

[25] Liu C, Zhang A Y, Pei D N, et al. Efficient electrochemical reduction of nitrobenzene by defect-engineered TiO_{2-x} single crystals[J]. Environ. Sci. Technol., 2016, 50: 5234-5242.

[26] Zhou W Y, Liu J Y, Song J Y, et al. Surface-electronic-state-modulated, single-crystalline {001} TiO_2 nanosheets for sensitive electrochemical sensing of heavy-metal ions[J]. Anal. Chem., 2017, 89: 3386-3394.

[27] Pei D N, Gong L, Zhang A Y, et al. Defective titanium dioxide single crystals exposed by high-energy {001} facets for efficient oxygen reduction[J]. Nat. Commun., 2015, 6: 8696.

[28] Liu G, Yang H G, Pan J, et al. Titanium dioxide crystals with tailored facets[J]. Chem. Rev., 2014, 114: 9559-9612.

[29] Liu S W, Yu J G, Jaroniec M. Anatase TiO_2 with dominant high-energy {001} facets: synthesis, properties, and applications[J]. Chem. Mater., 2011, 23: 4085-4093.

[30] Fang W Q, Gong X Q, Yang H G. On the unusual properties of anatase TiO_2 exposed by highly reactive facets[J]. J. Phys. Chem. Lett., 2011, 2: 725-734.

[31] Kuramitz H, Matsushita M, Tanaka S. Electrochemical removal of bisphenol A based on the anodic polymerization using a column type carbon fiber electrode[J]. Wat. Res., 2004, 38: 2331-2338.

[32] Quan X, Yang S G, Ruan X L, et al. Preparation of titania nanotubes and their environmental applications as electrode[J]. Environ. Sci. Technol., 2005, 39: 3770-3775.

[33] Koo M S, Cho K, Yoon J, et al. Photoelectrochemical degradation of organic compounds coupled with molecular hydrogen generation using electrochromic TiO_2 nanotube arrays[J]. Environ. Sci. Technol., 2017, 51: 6590-6598.

[34] Yuan S J, Sheng G P, Li W W, et al. Degradation of organic pollutants in a photoelectrocatalytic system enhanced by a microbial fuel cell[J]. Environ. Sci. Technol., 2010, 44: 5575-5580.

[35] Lee C Y, Bond A M. Evaluation of levels of defect sites present in highly ordered pyrolytic graphite electrodes using capacitive and faradaic current components derived simultaneously from large-amplitude fourier transformed ac voltammetric experiments[J]. Anal. Chem., 2009, 81: 584-594.

[36] Zhang J, Guo S X, Bond A M. Discrimination and evaluation of the effects of uncompensated resistance and slow electrode kinetics from the higher harmonic components of a fourier transformed large-amplitude alternating current voltammogram[J]. Anal. Chem., 2007, 79: 2276-2288.

[37] Wu M F, Jin Y N, Zhao G H, et al. Electrosorption-promoted photodegradation of opaque wastewater on a novel TiO_2/carbon aerogel electrode[J]. Environ. Sci. Technol., 2010, 44: 1780-1785.

[38] Yang Y, Li J X, Wang H, et al. An electrocatalytic membrane reactor with self-cleaning function for industrial wastewater treatment[J]. Angew. Chem., Int. Ed., 2011, 50: 2148-2150.

[39] Yang Y, Wang H, Li J, et al. Novel functionalized nano-TiO_2 loading electrocatalytic membrane for oily wastewater treatment[J]. Environ. Sci. Technol., 2012, 46: 6815-6821.

[40] Guo C, Ge M, Liu L, et al. Directed synthesis of mesoporous TiO_2 microspheres: catalysts and their photocatalysis for bisphenol A degradation [J]. Environ. Sci. Technol., 2009, 44: 419-425.

[41] Anglada A, Urtiaga A, Ortiz I. Pilot scale performance of the electro-oxidation of landfill leachate at boron-doped diamond anodes[J]. Environ. Sci. Technol., 2009, 43: 2035-2040.

[42] Yang H G, Sun C H, Qiao S Z, et al. Anatase TiO_2 single crystals with a

large percentage of reactive facets[J]. Nature, 2008, 453: 638-641.

[43] Yu J G, Low J X, Xiao W, et al. Enhanced photocatalytic CO_2-reduction activity of anatase TiO_2 by coexposed {001} and {101} facets[J]. J. Am. Chem. Soc., 2014, 136: 8839-8842.

[44] Xiang Q J, Yu J G, Jaroniec M. Tunable photocatalytic selectivity of TiO_2 films consisted of flower-like microspheres with exposed {001} facets[J]. Chem. Commun., 2011, 47: 4532-4534.

[45] Zhang D Q, Li G S, Yang X F, et al. A micrometer-size TiO_2 single-crystal photocatalyst with remarkable 80% level of reactive facets[J]. Chem. Commun., 2009, 45: 4381-4383.

[46] Yu J G, Dai G P, Xiang Q J, et al. Fabrication and enhanced visible-light photocatalytic activity of carbon self-doped TiO_2 sheets with exposed {001} facets[J]. J. Mater. Chem., 2011, 21: 1049-1057.

第 4 章

可见光辅助 TiO_{2-x} 单晶电催化降解污染物

4.1 概述

缺陷工程是一种通过调节中心金属晶格原子和电子结构以达到高活性结构自修饰的技术方法[1],该方法能够使过渡金属氧化物的电催化活性得到显著提高[1-5]。在能源和环境应用领域中,非化学计量的金属氧化物通常在O_2/CO_2还原、催化产氢和污染物降解方面表现出优于其本征氧化物的性能[5-8]。

相比于阴极还原法(详见第 7 章),阳极氧化是一种更有效、更广泛的水处理方法[9-11]。由于表面和亚表面氧空位以及其他缺陷活性位点在强阳极极化下的稳定性较差,非化学计量的过渡金属氧化物不能直接用于高偏压 EC 条件下的电极防污。例如,氢掺杂的 TiO_2 电极在长时间阳极氧化过程中会发生$\equiv Ti^{3+}$氧化和质子放电这两种副反应,导致其催化性能急剧下降[12]。因此,如何稳定非化学计量的金属氧化物上的氧空位缺陷和其他活性位点,对电化学水处理具有重要的意义。例如,Ti_4O_7 电极具有阳极氧化降解污染物的能力,如果存在阴极再活化过程,还原被氧化的$\equiv Ti^{3+}$并将质子插入重新排列的结构中,则其在电化学过程中是稳定的[2,13]。

光催化水处理是近三十年来被广泛研究的方向[14]。与强 EC 阳极氧化条件下的弱稳定性不同[12],氧空位在 PC 可见光照射下具有更好的稳定性[15-18],这说明 PC 可以起到稳定 EC 非化学计量金属氧化物表面和亚表面的缺陷活性位点(氧空位)的作用。因此,构建 PC(可见光)辅助的 EC 体系(即 PEC)是拓展水处理方法的一种新思路[19]。在该体系中,低偏压 EC 和可见光 PC 协同作用于单一的非化学计量金属氧化物上,EC 污染物的转化主要依赖于表面吸附后的直接电子转移机理[19],而·OH 介导的 PC 氧化可以矿化降解阳极表面的中间体聚合物,从而成为一种解决 EC 中电极失活问题的有效防污策略[20-27]。氧空位作为非化学计量金属氧化物表面和亚表面的缺陷型催化活性位点,可以在 PEC 体系中实现结构和催化性质的光化学稳定保护,以实现高效稳定的水处理过程[28-33]。

为了实现这一目标,需要解决光化学防护策略和 PEC 体系的关键问题——如何在单一电极上负载具有优异的 PC 和 EC 活性的双功能阳极材料[34-43]。TiO_2 由于其较宽的带隙(~3.2 eV),只能吸收和利用占大约 4.0% 太阳能的紫外光($\lambda < 378$ nm),难以利用能量较低、存在更为广泛的可见光[44-45],因此

通过表面缺陷化学来调节 TiO_2 的局部原子和电子结构,在保持优异 EC 活性的同时显著提升其可见光 PC 活性,成为一种非常有前途的研究设想[46-54]。

为了保护非化学计量金属氧化物表面和亚表面中作为缺陷活性位点的氧空位,同时拓宽光化学防污策略的适用范围,本章工作在上一章的研究基础上,尝试建立一个以 TiO_{2-x} 作为阳极催化剂的低偏压 EC 主导、PC(可见光)辅助的 PEC 体系来降解 BPA,同时深入分析该体系的降解效率、催化稳定性和能耗状况,考察表面和亚表面氧空位的结构和催化稳定性,验证缺陷活性位点介导的光化学保护机理,进一步拓展 PEC 体系在太阳光照条件下处理各类实际样品的应用潜力,开发出一种新型、高效、廉价、绿色、实用的水处理工艺。

4.2 TiO_{2-x} 单晶的可见光-电性能与光电耦合催化体系设计

4.2.1 TiO_{2-x} 单晶的制备

采用水热法制备了高能{001}暴露晶面的 TiO_2 SCs,然后在 H_2 气氛下,在 400 ℃下热还原为 TiO_{2-x} SCs(图 4.2 和图 4.4)[53]。对该 TiO_2 SCs 进行了形貌和结构的表征,并在常规三电极体系中测定了其 PC、EC 和 PEC 的性质。采用滴涂技术将 TiO_{2-x} SCs 沉积在碳纸基底上制备阳极电极。

4.2.2
TiO$_{2-x}$单晶的光电性能表征

在三电极体系中用 0.1 mol·L^{-1} Na$_2$SO$_4$ 水溶液作为支持电解质,以 TiO$_{2-x}$ SCs/碳纸作为工作电极,铂丝作为对电极,SCE 作为参比电极,通过光电流对光化学性能进行测试。光源为 500 W 氙灯(PLS-SXE500,Beijing Trusttech Co.,China)。入射单色光电转换效率(IPCE)采用 THORLABS 的 QE/IPCE 模块进行测量,使用 300 W 氙灯(Newport,model no. 6258)单色仪(Newport Cornerstone 130 1/8 m)。CV 扫描:以 0.1 mol·L^{-1} Na$_2$SO$_4$ 水溶液作为支持电解质、5.0 mmol·L^{-1} [Fe(CN)$_6$]$^{3-}$/[Fe(CN)$_6$]$^{4-}$ 作为氧化还原电对,扫速为 0.1 V·s^{-1},三电极体系。DPV 扫描:0.1 mol·L^{-1} Na$_2$SO$_4$,30 mg·L^{-1} BPA,分析体系的电化学性质。EIS 通过在 0.1 mol·L^{-1} Na$_2$SO$_4$ 水溶液中施加 5.0 mV 的交流电压、频率范围为 $10^5 \sim 10^{-2}$ Hz、外加电位和紫外光照进行测量。

4.2.3
双酚 A 在可见光和太阳光下的光电耦合催化降解

在一个圆柱形三电极单室电解池中进行 BPA 的催化降解。阳极有效面积为 6.0 cm^2,TiO$_{2-x}$ SCs 负载约为 0.30 mg,钛片作为阴极,SCE 作为参比电极,电极间距为 1.0 cm。使用电化学工作站(CHI 760D,Chenhua Co.,China)电解 80 mL 含 5.0~200.0 mg·L^{-1} BPA 与 0.1 mol·L^{-1} Na$_2$SO$_4$ 的水溶液,阳极偏压(SCE)范围为 0.5~2.0 V。使用高压 500 W 氙灯进行紫外光/可见光辐照,滤光片范围为 420 nm 以上(PLS-SXE500,Trusttech Co.,China)(图 4.1)。在 EC 和 PEC 体系中以本征态 TiO$_2$ 和缺陷型 TiO$_{2-x}$ 作为阳极进行污染物降解(未使用额外的外部电阻或 iR 补偿),所有测试均平行进行三次,计算出平均值与标准差。

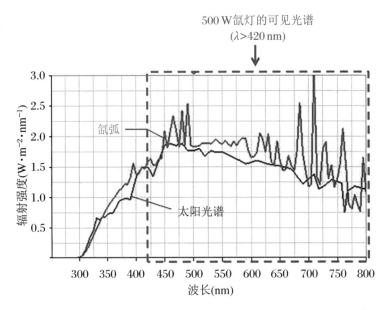

图 4.1　500 W 氙灯的可见光光谱与太阳光谱在 420～800 nm 波长范围的对比

红线为氙灯光谱，蓝线为太阳光谱

4.2.4
实际水样的处理

实际废水与地表水水样分别来自合肥当地某制药厂、印染工厂、巢湖和中国科学技术大学水上报告厅池塘。实际水样先用 0.45 μm 滤膜过滤，然后由 0.2 mol·L^{-1} Na$_2$SO$_4$ 等体积稀释，以保持足够的离子强度和电导率（0.1 mol·L^{-1} Na$_2$SO$_4$）。同时进行相同实验条件下的 EC 实验作为对照。所有测试都平行进行三次，计算出平均值与标准差。

4.2.5
电化学表征与辅助测试分析

使用光强计（UV751GD，Analytical Instrument Co.，China）测量太阳光光强。可见吸收光谱和荧光光谱分别采用紫外-可见分光光度计（UV-2401PC，

Shimadzu Co.，Japan)和荧光分光光度计(RF-5301PC，Shimadzu Co.，Japan)进行采集。BPA 的测定采用高效液相色谱法(HPLC-1100，Agilent Co.，USA)，采用 Hypersil-ODS 反相柱，VWD 检测器在 254 nm 处检测。流动相为水和甲醇的混合物(30∶70)，流速为 1.0 mL·min^{-1}。矿化效率由 TOC 分析仪(Vario TOC cube，Elementar Co.，Germany)来测定和计算。降解中间产物通过气相色谱-质谱联用仪(GCT Premier，Waters Inc.，USA)和液相色谱-质谱联用仪(LC-MS，LCMS-2010A，Shimadzu Co.，Japan)进行标定。

通过考察溶液在既定时间内的 TOC 变化，可以计算出催化反应体系的矿化电流效率与电化学平均能耗[2](参见公式(2.7)～(2.9))。

采用极限电流法计算了不同浓度 5.0～200.0 mg·L^{-1} BPA 在 TiO$_{2-x}$ 电极上的极限电流密度(表 4.1)，从而得到污染物在电极表面的传质系数(k_m，m·s^{-1})

$$k_m = \frac{j_{\lim}}{nFAC} \tag{4.1}$$

其中，$j_{\lim}$ 为极限电流(mA)，n 为 BPA 降解过程中的电子转移数，F 为法拉第常数，A 为有效电极面积(6.0 cm^2)，C 为 BPA 水溶液浓度(0.02～0.87 mmol·L^{-1})。

此外，针对 BPA 在缺陷型 TiO$_{2-x}$ 电极表面进行催化降解这类受吸附控制和完全不可逆的反应而言，可以通过 LSV 测试分析其阳极氧化峰电位(E_p)和扫速的对数函数($\log v$)之间的线性关系，进而计算在 EC 和 PEC 体系中的电子转移数(n)，其中 E_p 被定义为

$$E_p = E_0 + \frac{2.303RT}{\alpha nF}\lg\frac{RTk_0}{\alpha nF} + \frac{2.303RT}{\alpha nF}\lg v \tag{4.2}$$

式中，α 是传递系数，k_0 是标准的反应速率常数，n 为电子转移数，v 为扫速，E_0 为标准氧化还原电位，R 为气体常数，T 为绝对温度，F 为法拉第常数。

发现在 LSV 测量中阳极氧化峰电位(E_p)和扫速的对数函数($\log v$)之间的线性关系的斜率为 $2.303RT/(\alpha nF)$。因此得到在 EC 和 PEC 体系中 TiO$_{2-x}$ 电极表面 BPA 氧化的计算电子转移数约为 2.0(式(4.2))。进一步计算得到 BPA 在 TiO$_{2-x}$ 电极上的传质系数(k_m，m·s^{-1})与它们在 5.0～200.0 mg·L^{-1} 时的初始浓度相关，因而可以作为电解体系中的非均相阳极转化的有效度量指标(式(4.2))。

4.3

可见光辅助 TiO_{2-x} 单晶电催化降解双酚 A 的效能与机理分析

4.3.1

TiO_{2-x} 单晶的材料学特性

经过 H_2 高温煅烧还原后，TiO_2 的外观颜色变暗，而形貌和结构均保持稳定（图 4.2 和图 4.5）。ESR、XPS、Raman 和 DRS 表征结果均证实了表面氧空位的形成（图 4.4）[47-50]。

缺陷工程手段构建了非化学计量的原子和电子结构，使 TiO_2 的催化活性得到了显著提高（图 4.5）。TiO_{2-x} 电极在可见光下的光电流密度是本征态 TiO_2 的近 6 倍（图 4.5），表明其具有优越的载流子传输、分离和收集效率。TiO_{2-x} 的外部和内部量子效率（EQE 和 IQE）与 DRS 光谱非常接近，在 600 nm 时最大转换效率为 3.92%，在 420 nm 时最大转换效率为 2.26%（图 4.5(b)），均远远高于本征态 TiO_2。在 $[Fe(CN)_6]^{3-/4-}$ 处观测到一组清晰的氧化还原峰（图 4.5(c)），其中 TiO_{2-x} 电极的峰间距（ΔE_p）明显窄于其他对照组，这表明氧化还原电对与 TiO_{2-x} 电极之间的反应具有良好的可逆性，且由于氧化还原电对的表面和亚表面氧空位是反应的缺陷活性位点，氧化还原电对与 TiO_{2-x} 电极之间的电子转移更加容易。TiO_{2-x} 电极上的峰值电流大幅增加，这表明 BPA 氧化过程中的有效电极面积更大（图 4.5(d)）。BPA 氧化峰升高，电子转移电阻降低，可见光下·OH 生成能力增强，证实了 PEC 的协同效应（图 4.5(e~h)）。这些特性使 TiO_{2-x} 成为 PEC 中极好的阳极材料。

利用 TEM 和 HRTEM 证实材料为单晶结构（图 4.2），晶格条纹间距约为 0.19 nm，对应 {001} 晶面，选取区域电子衍射（SAED）作为 {001} 区域轴衍射图，表明上下方形面为 {001} 切面[44-45]。SAED 点衍射图显示出 TiO_{2-x} SCs 存在缺陷。TiO_{2-x} SCs 具有高结晶度的锐钛矿结构，这从 XRD 衍射图谱中得到了证实（图 4.2(a、b)）。

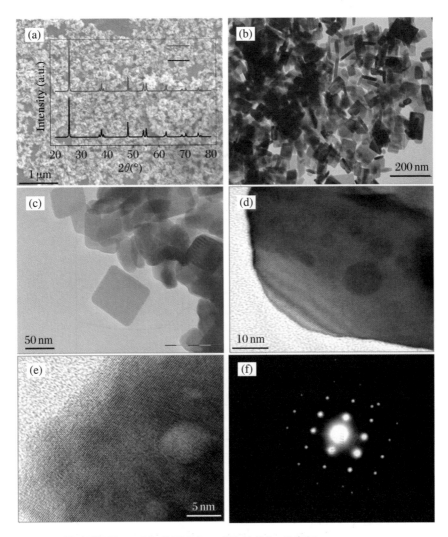

图 4.2　缺陷型 TiO_{2-x} SCs 的形貌(a~d)和结构(e,f)表征

图 4.3　TiO_{2-x} SCs 与 TiO_2 SCs 之间的颜色区别与相互转化

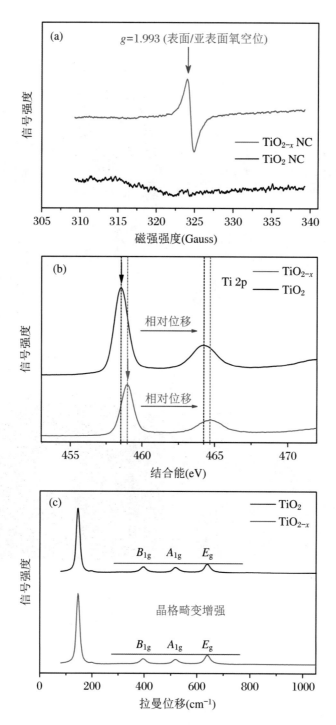

图 4.4　缺陷型 TiO_{2-x} SCs 中的氧空位缺陷表征：ESR(a)、Raman(b)、XPS(c)、DRS(d)、EIS(e) 和 Mott-Schottky 曲线(f)

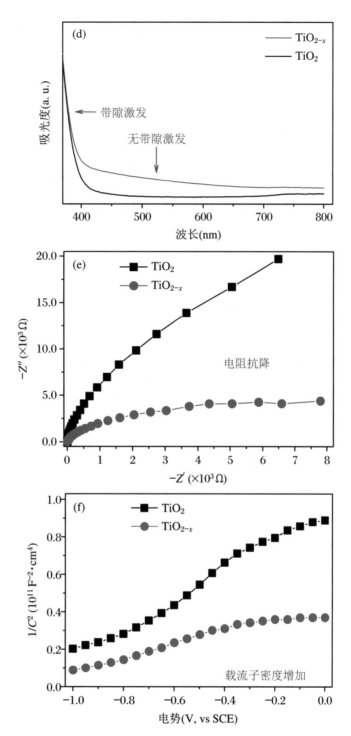

续图 4.4　缺陷型 TiO_{2-x} SCs 中的氧空位缺陷表征:ESR(a)、Raman(b)、XPS(c)、DRS(d)、EIS(e) 和 Mott-Schottky 曲线(f)

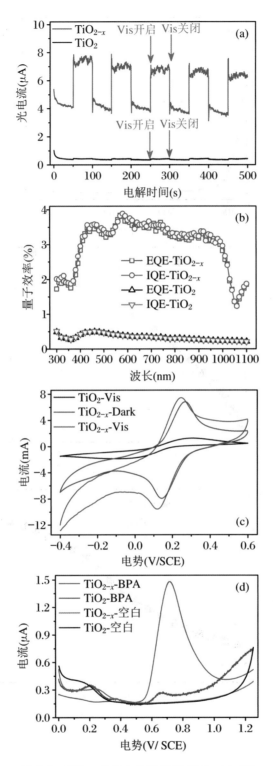

图 4.5 TiO$_{2-x}$ 的光电性能表征：PC(a、b)、EC(c、d) 和 PEC(e~h)

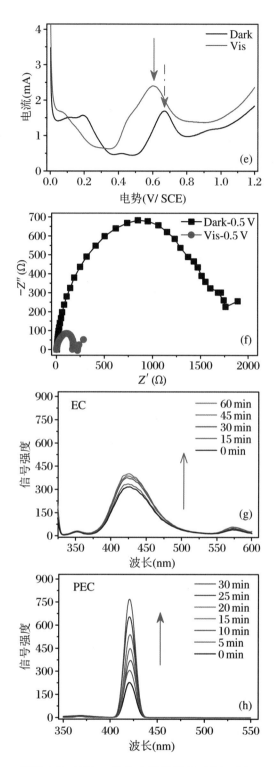

续图 4.5　TiO_{2-x} 的光电性能表征：PC(a、b)、EC(c、d)和 PEC(e～h)

为了确定≡Ti(Ⅲ)位点和氧空位的存在和位置,分别进行 ESR、XPS 和拉曼光谱测量(图4.4)[47-50]。在 ESR 图谱中观察到 $g=1.99$ 处的强信号,确认了缺陷≡Ti(Ⅲ)位点的存在(图4.4),对照组信号在 $g≈2.02$ 处,表明大部分≡Ti(Ⅲ)位点具有良好的化学稳定性[47-50]。此外,XPS 的结果证实了在 458.6 eV 的低结合能处存在着大量的≡Ti(Ⅲ)位点(图4.4(b))。此外,在缺陷型 TiO_{2-x} 的拉曼光谱中观察到一个清晰的峰加宽,表明氧空位数量增加以及缺陷的≡Ti(Ⅲ)位点的存在(图4.4(c))。TiO_2 和 TiO_{2-x} 的拉曼光谱差异较大,但 XRD 图谱差异不大,说明氧空位主要集中在表面和亚表面的≡Ti(Ⅲ)位点[47-50]。这一结论与 ESR 测试结果一致,说明 TiO_{2-x} 具有良好的环境应用前景,在循环实验中具有较高的结构和催化稳定性。

非化学计量的晶体结构中材料属性与缺陷/晶格氧密切相关。自掺杂≡Ti(Ⅲ)是一种扩展 TiO_2 可见光吸收的有效方法(图4.4(d)),并且不是通过降低其本征半导体带隙宽度实现的(3.2 eV)。这一现象主要是由于在价带顶部到导带底部的氧缺陷局域态下,以 $[Ov·Ti^{3+}]^+$ 的形式进行低能光子激发[8-10]。这个原位构筑的孤立亚能带在导带底部 0.3~0.8 eV 之间有不同的电子能级,从而具有广泛的可见光吸收,上述实验结果说明≡Ti(Ⅲ)主要来自于表面和亚表面的氧空位[8-10]。晶体缺陷对 TiO_{2-x} SCs 在多相催化过程中电子转移的导电性起着重要的促进作用(图4.4(e、f))。

4.3.2

双酚 A 降解效能评估

与 TiO_2 相比,TiO_{2-x} 电极对 BPA 的降解具有更强的 EC 活性(图4.6、图4.8、图4.14 和表4.1)。这些结果证实了结构/活性保护策略的开发对于利用缺陷型 TiO_{2-x} 作为阳极材料进行电化学水处理至关重要。TiO_{2-x} 的优异催化性能除了归结于单晶结构之外,主要归功于表面氧空位自掺杂与高能{001}晶面的设计策略[19,43,51,53],其优越的缺陷中心 EC 活性在 PEC 降解机理中发挥了重要作用(图4.6、图4.8、图4.14 和表4.1)。

当 PC 和 EC 耦合成 PEC 后,BPA 降解得到很大改善,说明其具有重要的催化优势(图4.6、图4.8、图4.14 和表4.1)。通常,在 +1.3 V/SCE 和可见光照射下,50.0 mg·L^{-1} BPA 在 2.0 h 完全降解,PEC 的速率常数 k_{PEC} 为 2.292 h^{-1},EC 的速率常数 k_{EC} 为 1.389 h^{-1},PC 的速率常数 k_{PC} 为 0.025 h^{-1},速率仅为

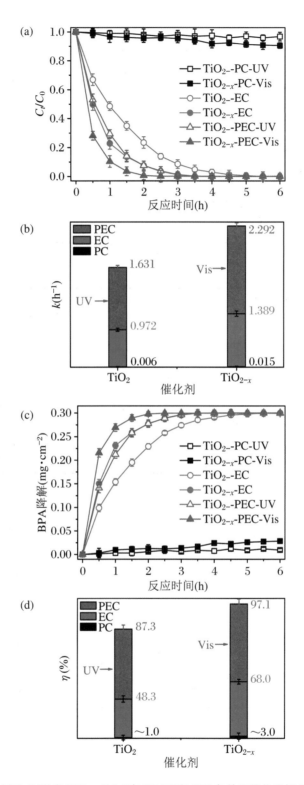

图 4.6　TiO_2(UV)和 TiO_{2-x}(Vis)在 PC、EC 和 PEC 条件下催化降解 BPA 的性能对比：降解速率(a)、反应速率(b)、转化率(c)和矿化率(d)

第4章

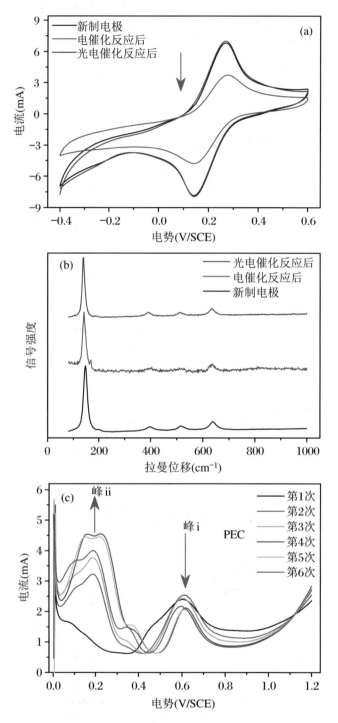

图 4.7　TiO_{2-x} 在 EC 和 PEC-Vis 体系下的防污性能对比：循环降解 BPA 前后的 CV(a)、Raman(b)、PEC(c)、EC(d)和 DPV(e、f)图谱

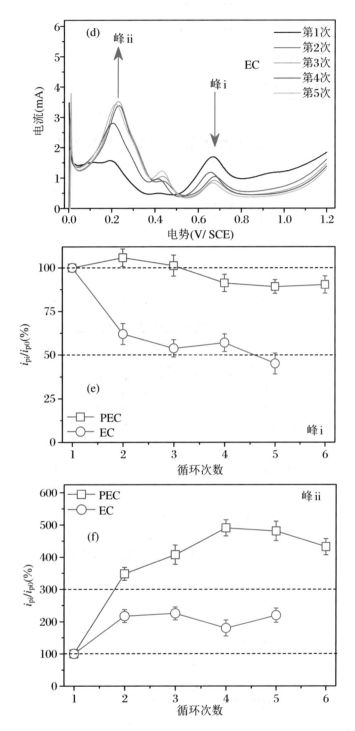

续图 4.7　TiO_{2-x} 在 EC 和 PEC-Vis 体系下的防污性能对比：循环降解 BPA 前后的 CV(a)、Raman(b)、PEC(c)、EC(d)和 DPV(e、f)图谱

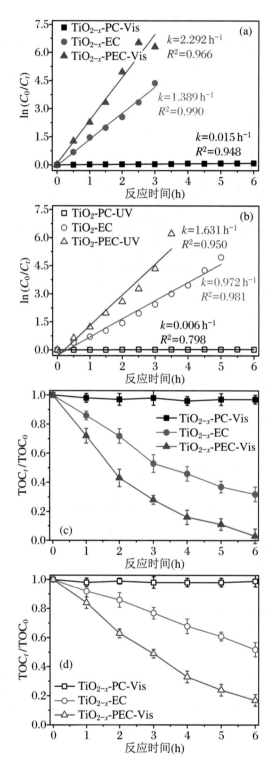

图 4.8 Vis 和 UV 条件下，缺陷型 TiO_{2-x}（a、c）和本征态 TiO_2（b、d）电极在 PC、EC 和 PEC 体系中对 BPA 的降解和矿化

表 4.1 TiO$_{2-x}$ SCs 电极在 PC,EC 和 PEC 条件下对 BPA 的降解实验参数①

	电势 (V/SCE)	浓度 (mg·L^{-1})	降解率 (%)②		一级动力学常数 (k_s, ×10^{-3} min)③		矿化率 (%)④		矿化电流效率 (%)⑤	
			TiO$_{2-x}$	TiO$_2$	TiO$_{2-x}$	TiO$_2$	TiO$_{2-x}$	TiO$_2$	TiO$_{2-x}$	TiO$_2$
PC	—	5	6.2±1.4	3.4±0.5	—	—	—	—	—	—
	—	20	1.6±0.3	1.2±0.4	—	—	—	—	—	—
	—	50	1.2±0.2	0.8±0.2	—	—	—	—	—	—
	—	80	0.4±0.1	0.4±0.3	—	—	—	—	—	—
	—	100	0.3±0.2	0.2±0.1	—	—	—	—	—	—
	—	150	2.0±0.2	0.7±0.3	—	—	—	—	—	—
	—	200	1.3±0.1	0.2±0.3	—	—	—	—	—	—
EC	0.5	30	33.72±1.5	16.77⑥	1.89±0.3	0.54⑥	19.32±1.2	14.23⑥	62.34±2.4	53.23⑥
	0.8	30	38.44±2.1	13.06⑥	3.34±0.4	0.50⑥	24.14±1.4	12.82⑥	50.80±3.1	46.60⑥
	1.0	30	48.73±2.3	20.74⑥	8.65±1.1	6.29⑥	42.22±2.1	18.14⑥	32.44±2.1	23.06⑥
		5	100.00±0.0	98.72⑥	49.90±2.1	35.57⑥	92.41±2.6	66.57⑥	2.02±0.4	3.98⑥
		20	100.00±0.0	62.87⑥	32.00±0.9	19.48⑥	97.52±2.2	40.50⑥	7.75±0.5	10.67⑥
		50	98.00±2.1	72.64⑥	28.72±0.8	6.23⑥	64.70±1.8	48.39⑥	40.02±1.8	32.51⑥
		80	90.65±0.9	40.07⑥	25.52±1.1	6.65⑥	70.20±1.7	24.34⑥	39.64±1.6	27.32⑥
		100	60.47±1.3	41.95⑥	14.06±1.0	5.17⑥	40.64±1.3	18.87⑥	38.23±1.4	25.57⑥
		150	53.18±1.5	30.40±2.1	11.85±0.8	1.12±0.2	37.15±1.5	14.32±0.7	32.41±0.9	17.59±1.0
		200	44.56±2.1	19.78±2.1	7.48±0.7	0.43±0.3	24.47±2.1	9.60±0.4	23.97±0.8	15.23±0.9
	1.5	30	100.00±0.2	97.89⑥	70.28±1.4	29.51⑥	95.02±2.8	74.52⑥	18.37±0.5	16.58⑥
	1.8	30	100.00±0.3	100.00⑥	89.03±1.6	36.86⑥	99.13±2.2	86.07⑥	13.35±1.0	11.39⑥

续表

电势 (V/SCE)		浓度 (mg·L^{-1})	降解率 (%)		一级动力学常数 (k_s, ×10^{-3} min)		矿化率 (%)		矿化电流效率 (%)	
			TiO$_{2-x}$	TiO$_2$	TiO$_{2-x}$	TiO$_2$	TiO$_{2-x}$	TiO$_2$	TiO$_{2-x}$	TiO$_2$
	0.5	30	34.57±1.4	26.40[6]	2.93±0.3	0.70[3]	30.34±1.7	24.65[3]	86.32±2.9	70.85[3]
	0.8	30	76.08±0.3	29.79[6]	20.33±1.0	2.00[3]	72.54±1.9	28.92[3]	81.17±3.2	71.36[3]
	1.0	30	89.55±2.1	42.28[6]	29.74±0.5	12.54[3]	78.45±2.4	34.71[3]	60.45±3.1	44.12[3]
		5	100.00±0.0	100.00[5]	82.98±2.6	47.75[5]	100.00±0.2	99.83[5]	4.87±0.2	5.72[5]
		20	100.00±0.0	98.25[5]	76.05±1.3	30.41[5]	94.35±3.1	89.87[5]	17.98±1.4	23.40[5]
		50	100.00±0.2	97.98[5]	42.62±1.4	23.99[5]	90.65±2.1	87.36[5]	40.98±2.0	52.52[5]
		80	100.00±0.1	80.83[5]	46.36±1.1	13.09[5]	92.22±1.9	72.32[5]	57.36±2.0	61.08[5]
PEC		100	94.24±0.9	72.17[5]	37.47±1.2	10.87[5]	81.81±1.6	58.04[5]	68.58±1.8	65.70[5]
		150	97.08±2.1	54.50±3.1	38.88±1.2	5.65±0.7	85.24±2.1	36.78±0.9	70.20±2.4	53.23±1.8
		200	98.35±0.8	43.68±2.4	30.60±1.0	3.90±0.2	80.90±3.0	26.74±0.4	64.37±2.3	40.08±2.1
	1.5	30	100.00±0.2	100.00[6]	85.85±2.4	26.62[6]	99.73±2.4	99.26[6]	20.42±1.4	24.79[6]
	1.8	30	100.00±0.3	100.00[6]	108.57±3.5	39.25[6]	99.20±2.7	99.83[6]	11.38±0.7	13.96[6]

注：① 平行三次。
② 反应时间为 3.0 h。
③ $\ln\left(\dfrac{C_0}{C_t}\right) = k_s \times t$。
④ 反应时间为 6.0 h。
⑤ MCE（矿化电流效率）$= \dfrac{\Delta TOC_{exp}}{\Delta TOC_{theor}} \times 100\%$。
⑥ 此前的相关研究数据（第 3 章）。
⑦ 浓度太低无法被准确检测。

~10.0%(图4.6(a~c))。这种协同效应也在矿化率结果中得到体现。PEC在6.0 h内降解了95.0% TOC以上,EC和PC分别降解了不到70.0%和3.0% TOC(图4.6(d))。PEC的优势可能是在PC的辅助下,高效生成和再生了更多的催化活性位点,在此基础上形成了高活性·OH_{free}[8-10,16-18]。PC中较低的BPA降解并不表明其活性差[19]。在可见光下,TiO_{2-x}所具有的表面和亚表面氧空位可以提高其PC活性(图4.6)。

为了验证TiO_{2-x}电极在PEC体系中的光化学保护作用,在BPA循环降解实验前后分别对其形貌、结构和电化学性能进行了测试(图4.4和图4.7)。使用过的EC电极的氧化电流远低于新电极的氧化电流,而PEC电极的氧化电流没有明显的下降(图4.7)。在142 cm^{-1}、394 cm^{-1}、512 cm^{-1}和634 cm^{-1}处发现了拉曼峰(图4.7(b)),没有发生增宽或位移,排除了TiO_{2-x}电极结构破坏的可能性。

在PEC中,BPA氧化峰相对稳定(图4.6、图4.7(c、e、f)),而在EC中则显著降低(图4.6、图4.7(c、e、f)),表明中间体聚合物的积累和TiO_{2-x}电极的污染。因此,原位引入PC可以为EC提供有效的保护,使其不污染电极表面,稳定TiO_{2-x}电极上的活性位点,极大地改善了EC占主导地位的PEC体系(图4.8、图4.14)。

进一步观察到EC对PC的促进保护作用(图4.4)。与BPA相似(图4.6),在相同的条件下与TiO_2电极相比,PEC对其他污染物的循环降解也比EC具有更高的活性和稳定性(图4.19)。

这些结果进一步证实PEC对TiO_{2-x}缺陷电极具有良好的光化学保护和电化学阳极防污能力(图4.6和图4.19)。BPA降解测试条件:溶液体积(80.0 mL),BPA浓度(5.0~200 mg·L^{-1}),Na_2SO_4浓度(0.1 mol·L^{-1}),阳极尺寸(6.0 cm^2),催化剂用量(0.05 mg·cm^{-2}×6.0 cm^2 = 0.30 mg),阴极尺寸(6.0 cm^2,钛片),电极间距(1.0 cm),偏压(+0.5~1.8 V/SCE),可见光(500 W氙灯,λ>420 nm),紫外光(500 W氙灯,λ<420 nm)、pH(~5.0)、温度(~20 ℃),搅拌速率(500 r·min^{-1}),反应时间(3.0/6.0 h)。

添加Na_2SO_4是为了提高溶液的导电性以达到降解防污的目的,因为在给定条件下,EC和PEC体系中的阳极极化和阴极极化都是惰性的(图4.6)。实验排除了Na_2SO_4与TiO_{2-x}电极之间的反应/相互作用(图4.6、图4.9、图4.10和图4.11)。

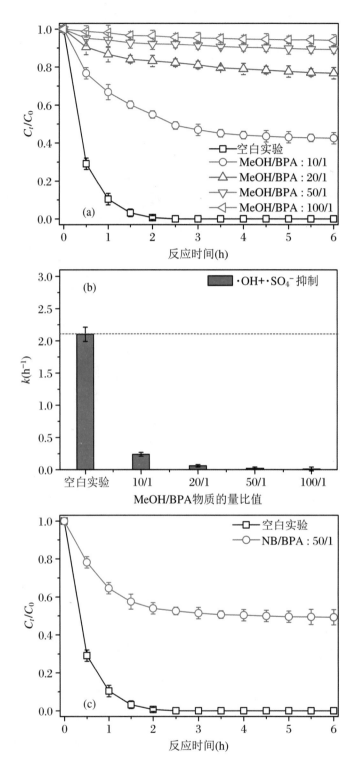

图 4.9　TiO_{2-x} 电极在 PEC 条件下催化降解 BPA 过程中的自由基抑制实验：·OH 和 ·SO_4^- 抑制（a、b）、仅 ·OH 抑制（c、d）和 $^1O_2$ 抑制（e、f）

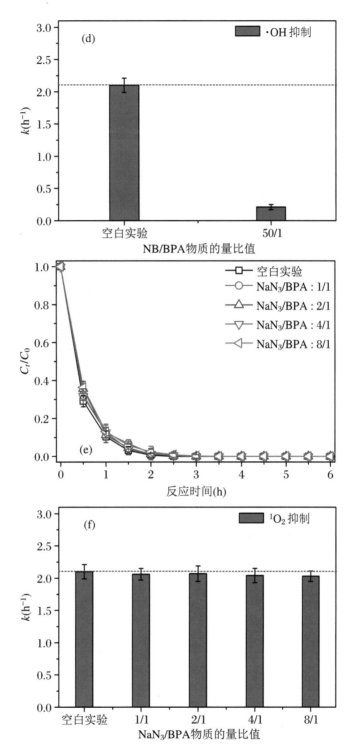

续图 4.9 TiO$_{2-x}$ 电极在 PEC 条件下催化降解 BPA 过程中的自由基抑制实验：·OH 和·SO$_4^-$ 抑制(a、b)、仅·OH 抑制(c、d)和 $^1O_2$ 抑制(e、f)

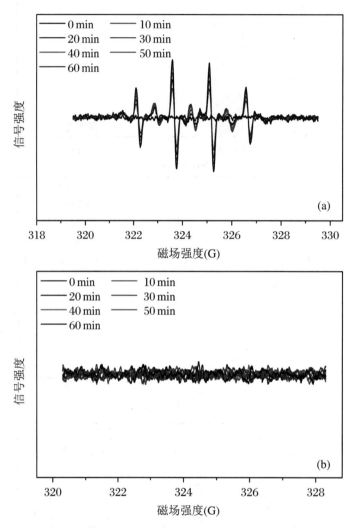

图 4.10 TiO$_{2-x}$ 电极在 PEC 体系中 DMPO-·OH(a)和 TEMP-^{1}O$_2$(b)的 ESR 信号

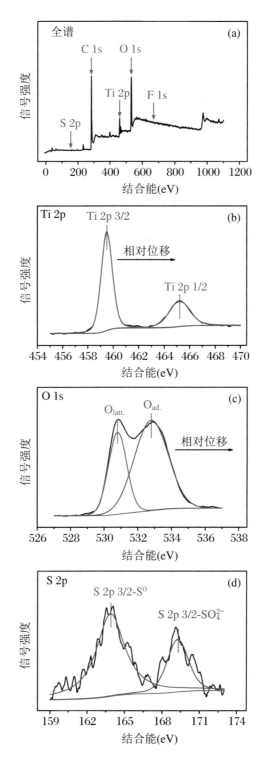

图 4.11　TiO_{2-x} 电极在 PEC 条件下循环 5 次降解 BPA 后的 XPS 谱图：全谱(a)、Ti 2p(b)、O 1s(c) 和 S 2p(d)

由于天然水体中的电解质浓度约为 $0.01\ \text{mol} \cdot \text{L}^{-1}$(图 4.6、图 4.18、图 4.20 和表 4.1),因此需要评估在 $0.01\ \text{mol} \cdot \text{L}^{-1}$ 天然电解质浓度下 PEC 的有效性。实验数据显示,该条件下的反应体系活性较低,动力学较慢,但是 PEC 明显优于 EC(图 4.12)。此外,BPA 和其他污染物的降解测试还存在光化学防护和防污能力(图 4.4)。由于电解质背景和溶液电导率对 PC 有效性的影响比 EC 要小(图 4.13),因此无论 Na_2SO_4 浓度如何,PC 在 PEC 体系中对 EC 的促进作用都可以维持(图 4.12 和图 4.13)。实验结果表明,支撑电解质对 PEC 体系处理效果非常重要。高剂量的 Na_2SO_4 保证了良好的效率和快速的动力学。此外,在处理后的水样中,无论电解质浓度如何,PEC 体系中的光化学保护和防污策略都可以稳定地对污染物进行降解。尽管天然水样的电解质浓度一般在 $0.01\ \text{mol} \cdot \text{L}^{-1}$ 左右,但 PC 辅助/EC 主导的 PEC 体系均体现出良好的活性和稳定性,具有广阔的实际应用前景。

通过测定 EC 和 PEC 体系中的极限电流来确定目标污染物 BPA 在缺陷型 TiO_{2-x} 电极上的传质系数(k_m, $\text{m} \cdot \text{s}^{-1}$)。结果表明,PEC 的 k_m 值($2.64 \times 10^{-5} \sim 2.48 \times 10^{-4}\ \text{m} \cdot \text{s}^{-1}$)高于 EC 的($2.33 \times 10^{-5} \sim 2.37 \times 10^{-4}\ \text{m} \cdot \text{s}^{-1}$)。这种差异可能是由于 TiO_{2-x} 电极在可见光照射下产生的焦耳热造成的(图 4.4)。该热量可以通过降低水的黏度来提高电极温度,促进传质。焦耳热效应提高了氧化速率,使反应动力学常数增大,从而降低了吸附速率。因此,焦耳热引起的氧化吸附平衡影响了去除效率。对比两种体系的计算值可以更恰当地解释缺陷型 TiO_{2-x} 电极所具备的 PEC 催化优势(图 4.4 和图 4.6)。PEC/EC 的优越性主要在于其独特的光化学保护和防污能力,以及可见光引入的焦耳热效应所导致的更高传质系数($\lambda > 420\ \text{nm}$)。

在可见光和紫外光照射下,分别用粉状 TiO_{2-x} SCs、TiO_2 SCs 和 P25 开展 $1.0\ \text{g} \cdot \text{L}^{-1}$ 催化剂投加量的三种 BPA 光催化降解实验。TiO_{2-x} SCs 在可见光照射下对 BPA 的降解和矿化均表现出良好的 PC 活性,但其降解和矿化效率低于紫外光照射下的 TiO_2 SCs(图 4.16)。TiO_{2-x} SCs 的优异的可见光驱动力使得 PC 活性得到很大程度的提高。同时,在 TiO_{2-x} 电极表面构筑独特的以 EC 主导的 PEC 协同催化机理。虽然在 $1.0\ \text{g} \cdot \text{L}^{-1}$ 的充足剂量下,TiO_{2-x} SCs 具有良好的 PC 活性(图 4.16),但仍远低于 PEC 体系(图 4.6(a))。这些结果表明,PC 和 EC 在 PEC 中的协同作用具有独特的 EC 主导的 PEC 机理(图 4.15、图 4.18 和图 4.20)。

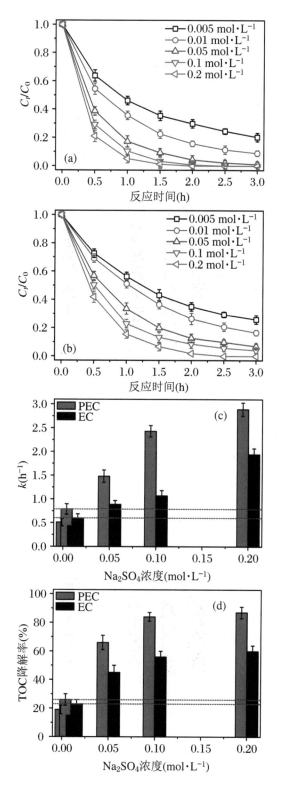

图 4.12 TiO$_{2-x}$ 电极在不同 Na$_2$SO$_4$ 电解质浓度下 PEC 催化降解 BPA：
(a) PEC 降解率；(b) EC 降解率；(c) 3.0 h 反应速率和 (d) 6.0 h 矿化率

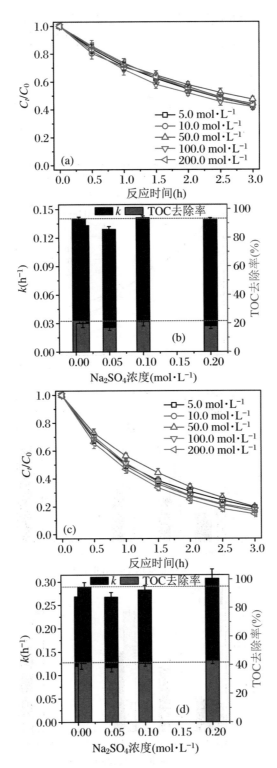

图 4.13 TiO$_{2-x}$、TiO$_2$ 和 P25 粉末在不同 Na$_2$SO$_4$ 浓度下对 BPA 的标准光催化 (UV/Vis) 去除实验：去除率(a、c、e、g)、反应速率和矿化率(b、d、f、h)

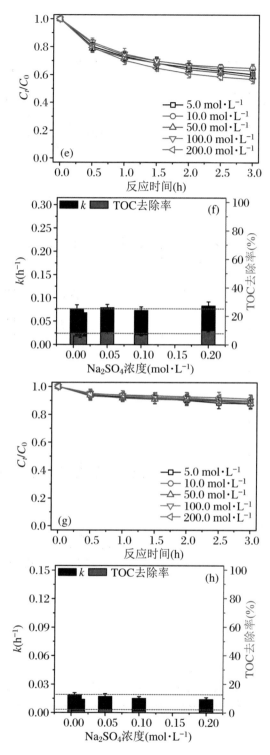

续图4.13 TiO_{2-x}、TiO_2 和 P25 粉末在不同 Na_2SO_4 浓度下对 BPA 的标准光催化(UV/Vis)去除实验:去除率(a、c、e、g)、反应速率和矿化率(b、d、f、h)

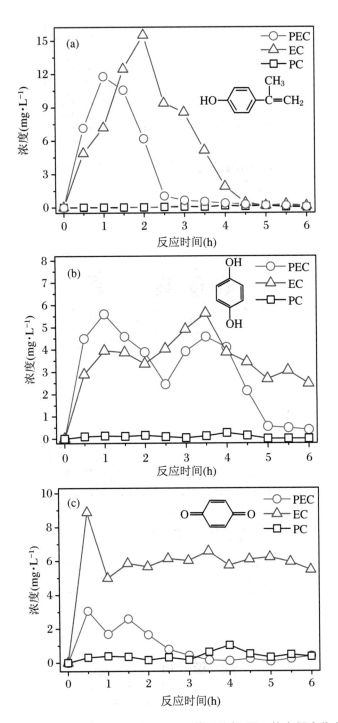

图 4.14 TiO$_{2-x}$ 电极在 PC、EC 和 PEC-Vis 体系降解 BPA 的中间产物分析：
(a) 4-异丙烯基酚；(b) 对苯二酚；(c) 苯醌；(d) 柠檬酸；(e) 马来酸；(f) 乙酸

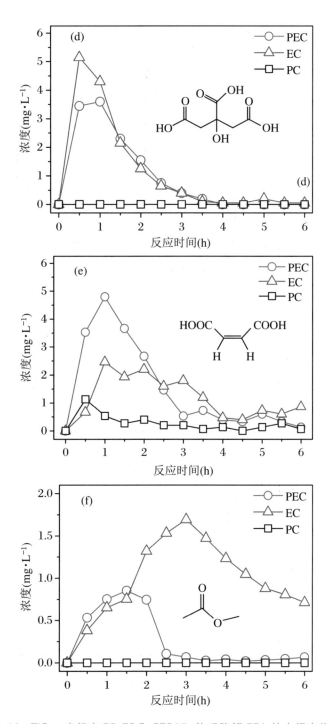

续图 4.14　TiO_{2-x} 电极在 PC、EC 和 PEC-Vis 体系降解 BPA 的中间产物分析：(a) 4-异丙烯基酚；(b) 对苯二酚；(c) 苯醌；(d) 柠檬酸；(e) 马来酸；(f) 乙酸

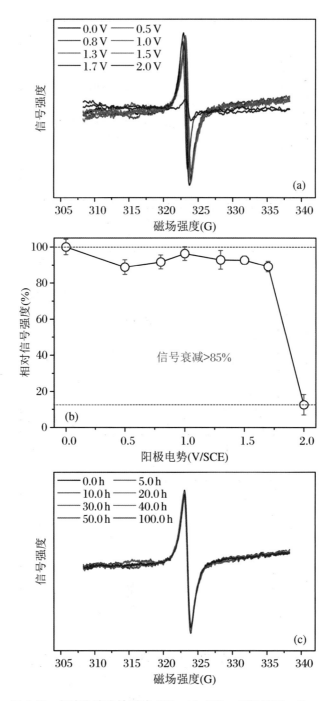

图 4.15 氧空位稳定性测试：EC(a、b)、PC(c、d) 和 PEC(e、f)

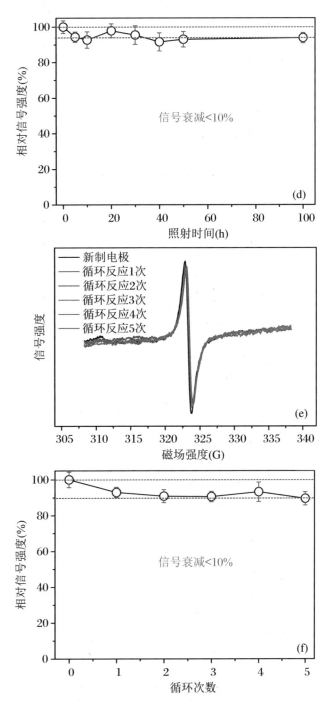

续图4.15　氧空位稳定性测试:EC(a、b)、PC(c、d)和PEC(e、f)

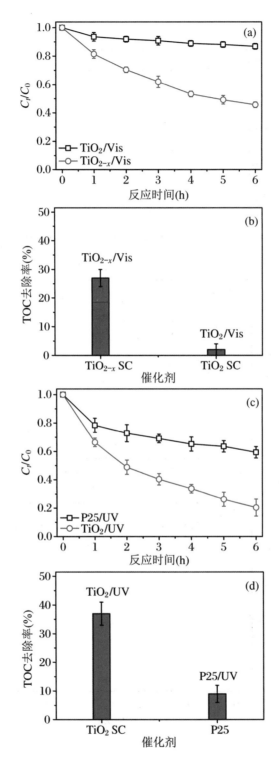

图 4.16 粉末态标准光催化降解污染物的性能对照：TiO_{2-x}/Vis、TiO_2/Vis(a、b)、TiO_2/UV 和 P25/UV(c、d)

为了进一步研究 TiO_{2-x} 介导的 BPA 在可见光照射下的降解(图 4.16),进行了空穴和·OH 的自由基抑制实验以及 ESR 测试[44-45]。实验结果揭示可见光照射下 TiO_2 和紫外光照射下 TiO_2 显著的·OH 介导的 BPA 降解机理(图 4.6)。虽然光生空穴只是反应性较小的种类,但所有的催化活性都主要归功于 TiO_2{001}暴露晶面的表面化学性质。PL 测试结果与实验数据高度一致(图 4.5(g、h))进一步说明,EC 生成聚集在阳极表面的 BPA 中间体聚合物,会因为 PEC 协同作用而发生高效的光化学降解矿化,即使 TiO_2 在可见光照射下的光化学活性低于紫外光照射下的(图 4.16),这种稳定、有效的防污机理仍然能在低偏压无析氧的电化学水处理体系中得到实际应用(图 4.4)。

4.3.3
双酚 A 降解中间产物与稳定性分析

通过 HPLC、LC-MS 和 GC-MS 等测试手段对 4-异丙烯基苯酚(图 4.14(a))、对苯二酚(图 4.14(b))、苯醌(图 4.14(c))、柠檬酸(图 4.14(d))、顺丁烯二酸(图 4.14(e))、乙酸(图 4.14(f))等主要降解中间产物进行了标定和定量。在相同时间内,EC 中 4 种中间产物的最大积累浓度均高于 PEC(图 4.14(a~c、e)),表明 PEC 体系降解 BPA 更快[19],EC 的其他两个中间产物的最大峰值累积浓度时间也高于 PEC(图 4.14(d、f)),这些结果与 TiO_{2-x} 电极在 PEC 体系中具有比 EC 更好的 BPA 降解性能高度一致(图 4.6 和表 4.1)。

我们通过 5 次循环降解实验研究了 PEC 在 TiO_{2-x} 电极上的催化稳定性。在 5 个周期内,BPA 的降解和矿化效率几乎没有降低(低于 5.0%),且始终显著高于 EC,说明 PEC 在长期使用过程中具有良好的催化稳定性。相比之下,EC 的降解虽然相对稳定,但矿化程度却在不断下降,这可能是污染物中间产物的吸附积累导致的电极失活(图 4.15)。因此,在 EC 和 PEC 中,TiO_{2-x} 电极的防污性能差异较大,这也得到了其他结果的证实。EC 和 PEC 循环实验均未出现峰增宽和位移,因此在污染物降解的长期使用中,TiO_{2-x} 电极未发生结构破坏(图 4.15(d))。

4.3.4
光化学保护下氧空位的结构稳定性

氧空位作为表面和亚表面的催化活性中心,其结构稳定性在催化反应中至关重要[4-7]。在 EC 体系中,当外加偏压低于 +1.7 V 时,TiO_{2-x} 电极的表面氧空位浓度在经历长周期阳极氧化后仍基本保持不变(图 4.15(a、b))。然而,这个阈值电压从热力学上不足以通过 ·OH 介导的氧化路径实现典型有机污染物的有效矿化($E^0(H_2O/·OH) = 2.80$ eV, $E^0(H_2O/O_2) = 1.23$ eV)。因此,与异质掺杂不同,由于表面和亚表面氧空位的结构敏感性,自掺杂的 TiO_{2-x} 电极不能稳定地用于高阳极偏压(如 2.0 V)下的电化学水处理,尽管它表现出比原始 TiO_2 更高的 EC 活性(图 4.5、图 4.6 和表 4.1)。除了表面和亚表面氧空位活性位点的破坏性氧化,TiO_{2-x} 在高偏压阳极氧化过程中也可能存在其他失活方式,如 H 原子的析出和缺陷型 ≡Ti^{3+} 位点的氧化($H^+ + Ti^{(4+)}O_2 \rightarrow Ti^{(3+)}O_2-H^+ \leftrightarrow Ti^{(4+)}O_2 + H^+ + e^-$)[5,13]。

在 PC(Vis)体系中,当反应时间延长到 100 h 以上(图 4.15(c、d))时,表面氧空位依旧表现出良好的结构稳定性($\lambda > 420$ nm)。相比高于 +1.7 V 的电化学阳极极化,可见光的光化学活化可以很好地维持氧空位(作为氧化还原循环中表面和亚表面的反应活性位点)的缺陷态原子和电子结构(图 4.17)。

在 PEC 体系中,TiO_{2-x} 电极的氧空位在阳极极化和可见光活化的长期催化过程中,也表现出良好的结构稳定性(图 4.18(e、f))。经过 5 次循环降解实验,ESR 信号保持高度稳定(图 4.18(e、f)),这一结果很好地解释了 PEC 体系对 BPA 具有高效稳定降解矿化的原因。由于表面和亚表面氧空位在低阳极偏压和可见光照射下具有良好的结构和催化稳定性,而在高阳极偏压下稳定性较差(如 +2.0 V(图 4.15(a~d))),PEC 的催化协同机理进一步验证了 TiO_{2-x} 的缺陷活性(图 4.4)。

4.3.5
缺陷活性位点介导的光化学保护机理

在 PEC 中，原位耦合的 PC 可以在可见光下矿化中间体聚合物，提高 TOC 的去除效果，并消除电极表面的污染物(图 4.6、图 4.16 和图 4.19)。它还可以稳定氧空位作为 TiO_{2-x} 电极上的缺陷活性位点，用于可持续催化(图 4.5)。TiO_{2-x} 通过四种非带隙激发路径(反应(4.3)~反应(4.5)、图 4.2 和图 4.4)同时被阳极偏压和可见光活化。在 PEC 中 TiO_{2-x} 表面的中心金属晶格，以及氧空位的结构和催化稳定性由 EC 和 PC 的氧化还原循环机理控制(反应(4.3)~反应(4.9)、图 4.2 和图 4.4)[11-19,43]。

$$TiO_{2-x} + 外加偏压 \rightarrow TiO_{2-x}(h^+)_{DB} + e^-_{DB}(缺陷态亚能带激发) \quad (4.3)$$

$$TiO_{2-x} + 外加偏压 \rightarrow TiO_{2-x}(h^+)_{CB} + e^-_{CB}(导带激发) \quad (4.4)$$

$$TiO_{2-x} + 可见光 \rightarrow TiO_{2-x}(h^+)_{VB} + e^-_{DB}(价带激发) \quad (4.5)$$

$$TiO_{2-x} + 可见光 \rightarrow TiO_{2-x}(h^+)_{DB} + e^-_{CB}(缺陷态亚能带激发) \quad (4.6)$$

$$TiO_{2-x}(h^+)_{DB} + \equiv C-OH \rightarrow \equiv C-O \cdot \rightarrow \equiv C-O \cdot \cdot O-C \equiv \rightarrow 聚合产物 \quad (4.7)$$

$$TiO_{2-x}(h^+)_{CB} + \equiv C-OH \rightarrow \equiv C-O \cdot \rightarrow \equiv C-O \cdot \cdot O-C \equiv \rightarrow 聚合产物 \quad (4.8)$$

$$TiO_{2-x}(\cdot OH)_{bound} + \equiv C-OH \rightarrow \equiv C-O \cdot \rightarrow \equiv C-O \cdot \cdot O-C \equiv \rightarrow 聚合产物 \quad (4.9)$$

$$TiO_{2-x}(h^+)_{VB} + 聚合产物 \rightarrow 开环中间体 \rightarrow CO_2 + H_2O \quad (4.10)$$

$$TiO_{2-x}(\cdot OH)_{free} + 聚合产物 \rightarrow 开环中间体 \rightarrow CO_2 + H_2O \quad (4.11)$$

$$TiO_{2-x}(\cdot OH)_{bound}/TiO_{2-x}(\cdot OH)_{free} \rightarrow 1/2O_2 \uparrow + H^+ + e^- \quad (4.12)$$

在 EC 作用下(E = +1.3 V/SCE)，缺陷态亚能带被直接活化，产生电子 e^-_{DB} 到 TiO_{2-x} 表面(反应 4.3)[11,19,43]。该氧空位通过两种不同的非带隙激发机理被转移到氧化态，一种是缺陷态亚能带，另一种是导带(反应(4.3)和(4.4)、图 4.2(a)和图 4.4)。然而，一方面，这种直接 EC 活化在高阳极偏压条件下可能会破坏表面和亚表面氧空位的原子和电子结构；另一方面，留在缺陷态亚能带内的电生低活性 h_{DB}^+ 不能很好地捕获污染物的电子，也不能有效地在再生缺陷活性位点进行持续反应(氧空位不能快速地从氧化态返回基态)(反应(4.3)、图 4.2 和

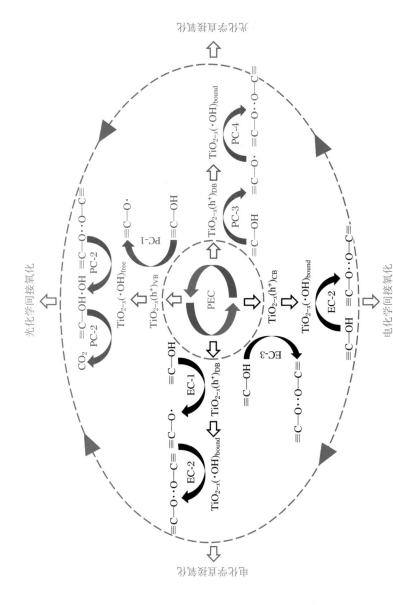

图4.17 PEC-Vis-TiO$_{2-x}$体系的缺陷活性中心光化学保护机制反应路径

EC-1: 缺陷带隙空穴介导的电化学直接氧化,TiO$_{2-x}$(h$^+$)$_{DB}$;EC-2: 表面结合态·OH介导的电化学间接氧化,TiO$_{2-x}$(·OH)$_{bound}$;EC-3: 导带空穴介导的电化学直接氧化,TiO$_{2-x}$(h$^+$)$_{CB}$;PC-1: 价带空穴介导的电化学直接氧化,TiO$_{2-x}$(h$^+$)$_{VB}$;PC-2: 游离态·OH介导的光化学间接氧化,TiO$_{2-x}$(·OH)$_{free}$;PC-3: 缺陷带隙空穴介导的光化学直接氧化,TiO$_{2-x}$(h$^+$)$_{DB}$;PC-4: 表面结合态·OH介导的光化学间接氧化,TiO$_{2-x}$(·OH)$_{bound}$

图4.4)。虽然 h_{DB}^+ 电生空穴的催化活性仍在缺陷态亚能带可以增加与外部的偏压(获得污染物电子的能力),但当阳极偏压高于 +1.7 V/SCE 时,表面缺陷活性位点因为原子和电子结构遭到强烈破坏而无法进行再生(图4.15(a、b))。因此,EC 仅在稳定性较差的缺陷型 TiO_{2-x} 电极上发生了部分转化(反应(4.10)~反应(4.12)、EC-1、EC-2、EC-3、图4.2、图4.4 和图4.6)。

氧化型 $TiO_{2-x}(h^+)_{DB}$ 和氧化型 $TiO_{2-x}(\cdot OH)$ 是外加偏压和可见光条件下电极表面上的两种主要活性物种,通过 $TiO_{2-x}(h^+)_{DB}$ 电化学氧化生成的中间体聚合物被 $TiO_{2-x}(\cdot OH)$ 光化学矿化,因而使 PEC 体系双功能电极上的污染物全部去除。

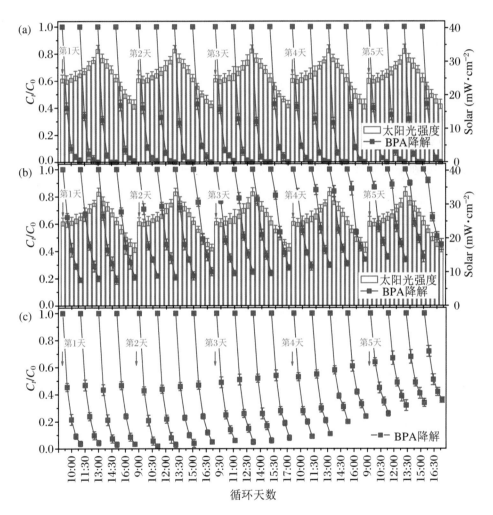

图4.18 污染物降解循环稳定性实验:(a) 太阳光-EC-TiO_{2-x};(b) 太阳光-EC-TiO_2;(c) EC-TiO_{2-x}

在 PC 活化中($\lambda > 420$ nm),缺陷态亚能带直接活化生成电子 e_{DB}^-,然后传递到 TiO_{2-x} 表面(反应(4.5)、图4.4 和图4.2(b))[15-18]。价带通过两种不同的非带隙激发机理转变为氧化态,一种是价带,另一种是缺陷态亚能带(反应(4.5)和反应(4.6)、图4.2(b)和图4.4)。与EC活化(反应(4.3))不同,缺陷态亚能带并没有被大量直接活化(仍然主要处于基态)(反应(4.5)、图4.2(b)和图4.17)。在PC活化过程中,缺陷态亚能带仅作为价带到TiO_{2-x}表面的电子媒介(反应(4.5)和图4.2(b))[47-50]。价带中生成的高活性h_{VB}^+可以很容易地从污染物中夺取电子,实现再生并保护表面缺陷活性位点,使其发生持续的催化反应(价带可以很容易地从氧化态返回基态)(图4.4和图4.15(c、d))。因此,PEC能发生良好稳定而彻底的污染物矿化反应(图4.8(e)和(f)和图4.7),因为PC通过非本征禁带宽度激发可以保护缺陷态亚能带、稳定TiO_{2-x}电极表面的缺陷活性位点以及定向调控析氧副反应(反应(4.10)~反应(4.12)、图4.4中的PC-1、PC-2、PC-3和PC-4)。

在TiO_{2-x}电极上,光化学保护策略不仅可以防止阳极电极被污染,而且可以稳定作为缺陷活性位点的表面和亚表面氧空位,实现稳定的电化学去污。光化学保护策略可以有效地强化TiO_{2-x}电极上的催化性能(图4.4)。TiO_{2-x}的氧空位缺陷在可见光照射下能有效生成·OH(图4.5),这可以很好地解释其在电化学水处理中的阳极聚合产物矿化能力,以及PEC体系中的光化学保护能力(图4.7和图4.15)。虽然在紫外光照射下TiO_2的防污能力远远大于暗态条件,仍低于在可见光照射下TiO_{2-x}的催化能力,但从DPV图谱的峰值变化能看出,PEC比PC具有更高的活性(图4.7)。与相同条件下的TiO_2电极相比,TiO_{2-x}电极上DPV反应更高、更稳定,BPA转化生成的中间产物更多(图4.7)。

TiO_{2-x} SCs的较好的EC活性可能主要来源于晶体缺陷的特殊原子电子结构(图4.4、图4.5、图4.15和图4.20)[47-50]。邻近氧空位的电子离域化,活化金属中心,对TiO_{2-x} SCs上的环境污染物的吸附、极化和转化反应更强烈(图4.2)。一方面,在TiO_{2-x}富含缺陷态原子构型的晶格结构上,其表面和亚表面存在大量可触及的活性位点、缺陷态≡Ti(Ⅲ)和氧空位(图4.2(e)和图4.4(a)~(c)),可以促进晶格结构改性并形成额外的缺陷活性位点。另一方面,作为内在催化反应活性位点的≡Ti(Ⅲ)中心和表面氧空位具有更强的电子交换能力(图4.5)。第三方面,缺陷子能带可以改善载流子密度和电导率,使得表面活性位点间的电子转移阻力大幅降低,这可以通过非带隙电子激发机理下大于400 nm的非特征的宽吸收光谱来直接说明(图4.2和图4.4(f))。

TiO_{2-x}电极在可见光下具有良好的光化学保护能力,这可能是因为其表面和亚表面上Ti晶体缺陷调制的局部原子和电子结构改善了EC的活性(图

4.17)$^{[47-50]}$。本研究首次提出协同 PEC 体系中 TiO$_{2-x}$ 电极上 EC 促进 PC 保护机理(图 4.17)。EC 活性强,且具有较强的电化学活化能力,可以促进污染电极表面中间体聚合物的光化学降解。缺陷型≡Ti(Ⅲ)晶格位点可以作为活化 TiO$_{2-x}$ 表面中间体聚合物电荷转移的高效电子穿梭体,构建"EC 促进 PC 保护"的独特电极抗污染模型。与紫外光照射下 TiO$_2$ 电极相比,可见光照射下 TiO$_2$ 电极上的 PC 矿化更容易去除 EC 转化的中间产物。TiO$_{2-x}$ 电极表面的中间体聚合物在阳极极化条件下因其强大的 EC 活性而被热力学活化,从而在光化学保护过程中实现能量势垒的降低和反应动力学的加速(图 4.2 和图 4.4)。TiO$_{2-x}$ 电极优秀的 EC 活性也可以减少污染物电氧化所需要的外部偏压,而弱化的阳极偏压和极化强度进一步有利于表面和亚表面氧空位缺陷活性位点的光化学保护(图 4.1 和图 4.8)。此外,与 UV 照射下 TiO$_2$ 电极上苛刻的带隙激发通路相比(图 4.2),自掺杂的缺陷型 TiO$_{2-x}$ 亚能带结构可以实现可见光照射下温和的光化学活化非本征禁带宽度激发路径。因此,PEC 体系对 TiO$_{2-x}$ 电极表面的防污和活性位点的稳定具有高效的 PC 保护能力。

4.3.6
反应机理及双酚 A 降解路径分析

在 TiO$_{2-x}$ 电极上 EC 主导的 PEC 体系中,BPA 降解主要由 EC 引发,部分由 PC 引发(图 4.19)。

在 EC 体系中,BPA 的降解去除主要通过由电生空穴、缺陷能带空穴(h_{DB}^+)和导带空穴(h_{CB}^+)主导的直接氧化机理(路径 1)。在 +1.3 V 偏压条件下(图 4.6),电生空穴在缺陷能带和导带都有保留,因此只能氧化吸附水生成低活性表面吸附态·OH$_{bound}$ 间接进行转换。

BPA 首先转化为 4-异丙基苯酚和苯酚;随后,它们在直接路径中被进一步氧化为脂肪酸;最后,脂肪酸部分矿化$^{[19,43]}$。许多顽固的聚合产物堆积在电极表面,阻碍活性位点发生进一步反应,导致电极失活严重,性能下降。

在 PC 体系中,BPA 的降解是通过价带和缺陷态亚能带、h_{VB}^+ 和 h_{DB}^+(路径 1)中残留的光生空穴直接机理和高活性游离态·OH$_{free}$(路径 2)的间接氧化机理进行的$^{[47-50]}$。PC-1 因其对 TiO$_{2-x}$ 表面的强吸水性和光生空穴的高能量而难以进行;PC-2 主要通过"游离态·OH"在固液界面中进行,"游离态·OH"定位于 TiO$_{2-x}$ 表面,不含 TiO$_{2-x}$(·OH)$^{[47-50]}$。

▷ 第4章

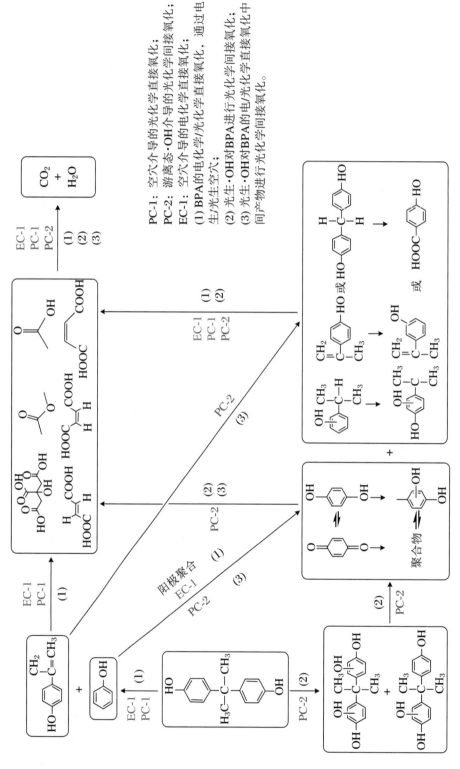

图4.19 BPA在PEC-Vis-TiO$_{2-x}$体系中的可能降解路径(+1.3 V/SCE)

PC-1：空穴介导的光化学直接氧化；
PC-2：游离态·OH介导的光化学间接氧化；
EC-1：(1) BPA的电化学/光化学直接氧化，通过电生光生空穴；
(2) 光生·OH对BPA进行光化学间接氧化；
(3) 光生·OH对BPA的电/光化学直接氧化中间产物进行光化学间接氧化。

144

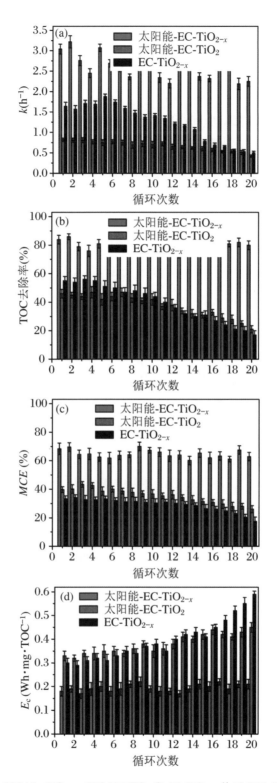

图 4.20 PEC-Vis-TiO$_{2-x}$、PEC-UV-TiO$_2$ 和 EC-TiO$_{2-x}$ 体系 BPA 降解循环实验的反应速率(a)、矿化率(b)、电流效率(c)和能耗(d)变化趋势

间接·OH_{free}介导的氧化过程也主要由三个步骤（路径 2）组成。这些中间产物不稳定，易通过异丙烯桥的裂解分解为单环芳香族化合物；随后，初始形成的芳香族化合物经过进一步的环裂解形成脂肪酸；最后，这些有机酸被进一步氧化成 CO_2，实现了 BPA 的彻底矿化。

在 PEC 中，BPA 降解中间产物由路径 1 中 EC、PC 的电生、光生空穴进行直接氧化，通过 PC 所产生的·OH_{free} 间接矿化（路径 3）。此外，形成的中间体聚合物在 EC 直接转换路径 1 也可以氧化为脂肪酸，最后通过 PC 间接机理所产生的·OH_{free} 矿化（路径 3）。因此，释放 TiO_{2-x} 表面活性位点，对失活电极进行再生，才能从根本上提升污染物降解速率。总而言之，这两种不同的催化过程可以协同加速整体的降解动力学，这可能很好地解释了 TiO_{2-x} 电极上 PEC 体系中 BPA 的良好降解效果（图 4.6、图 4.18 和图 4.20）。

4.3.7
耦合体系在太阳光照射下的处理性能

为了进一步评估 PEC 体系的光化学防护能力，在实际太阳光照射下进行测试（图 4.18）。PEC-TiO_{2-x} 体系中 BPA 去除与太阳光强度密切相关（图 4.18(a)），其效果优于 EC-TiO_{2-x} 和 PEC-TiO_2 体系（图 4.18(b)、(c)）。中午（11：00—13：00）和下午（13：00—15：00）的 BPA 降解率高于上午（9：00—11：00）和傍晚（15：00—17：00)（图 4.18(a)）。PEC-TiO_{2-x} 体系在 5 天内进行了 20 次循环降解实验，具有更高的活性和稳定性（图 4.20）。这些结果表明，TiO_{2-x} 电极在真实的阳光照射下具有良好的光化学保护能力。矿化电流效率（MCE）较高，平均能耗（E_a）较低（图 4.20(a)）。

以下进一步探讨和验证 PEC 体系在实际光照（太阳光）条件下处理实际样品的可行性。与 EC 体系相比，在 PEC 体系中，即使在天然电解质背景下，也能获得更高的效率和更快的动力学（图 4.21）。这些结果表明，TiO_{2-x} 电极可以在电化学水处理中获得太阳能，提高效率，降低成本。这种光化学保护策略使 TiO_{2-x} 电极具有电化学水处理的阳极材料的优点。需要指出，在某些地表水样品的降解过程中 PEC 对 EC 的增强相对较低（图 4.4 和图 4.21(h)），这些结果对于 PEC 体系不具有足够的指导意义，因为出现实验中现象的最主要原因可能是由于水样有机污染物浓度较低所致（约为 19.0 mg·L^{-1} TOC，图 4.4 和图 4.21(h)）。对于本研究中 PEC 和 EC 等在水氧化前低偏压的表

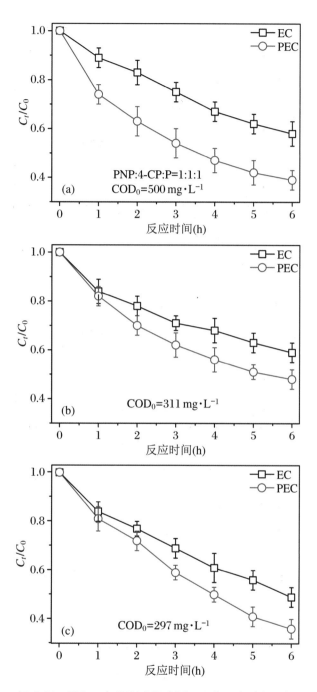

图 4.21 TiO$_{2-x}$ 在 PEC(太阳光)和 EC 体系中对实际水样的处理：多酚混合废水(a)、垃圾渗滤液(b)、印染废水(c、d)、腐殖质水样(e、f)和地表水(g、h)

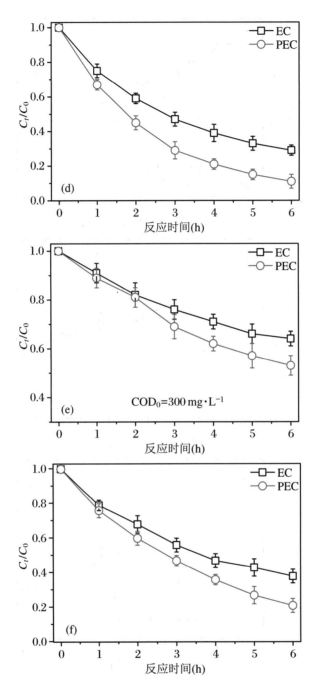

续图 4.21　TiO_{2-x} 在 PEC(太阳光)和 EC 体系中对实际水样的处理：多酚混合废水(a)、垃圾渗滤液(b)、印染废水(c、d)、腐殖质水样(e、f)和地表水(g、h)

面介导的污染物降解，直接阳极氧化是一种非均相电化学反应，其总速率电阻

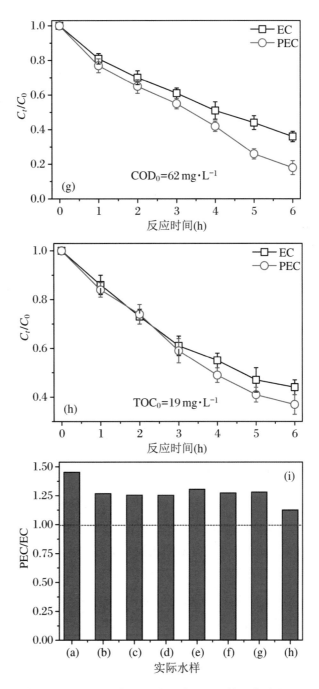

续图 4.21 TiO$_{2-x}$ 在 PEC(太阳光)和 EC 体系中对实际水样的处理:多酚混合废水(a)、垃圾渗滤液(b)、印染废水(c、d)、腐殖质水样(e、f)和地表水(g、h)

是传质电阻和化学反应电阻的总和[1-3]。电子从基底直接转移到电极只发生在基底被预先吸附到电极表面时,而传质通常是整个反应的主要限速步骤(图4.22)。

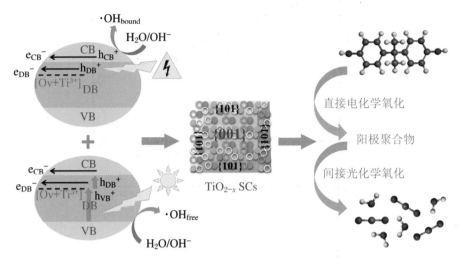

图 4.22　TiO_{2-x} 可见光辅助电催化降解 BPA 示意图

本章小结

基于上一章的研究内容,本章工作从表面和亚表面富含氧空位的缺陷型 TiO_{2-x} 单晶出发,将低压电催化与可见光催化进行协同耦合,研发了以电催化主导的可见光耦合电催化氧化体系。利用污染物降解实验、光电化学表征、自由基捕获和晶体结构分析等实验结果,揭示了缺陷型 TiO_2 单晶介导的光电耦合体系的氧化效能和作用机理。研究发现,在污染物光电催化降解过程中,作为催化活性位点的表面和亚表面氧空位通过可见光非带隙激发路径得到了有效保护,因而可以保持良好的结构和催化稳定性。进一步将电化学水处理与可见光辐照相耦合,发展了游离态·OH 介导的光化学阳极防污新策略,该体系在太阳光照射下能够高效、稳定降解污染物和处理废水。可见光耦合电催化不仅解决了电极污染问题,而且保护了缺陷型电极材料表面的催化活性位点,拓宽了光电催化体系的光吸收范围,显著提升了低偏压条件下 TiO_{2-x} 的电催化降解效能,为污染物的高效降解去除提供了新的技术。

参考文献

[1] Chen D J, Chen C, Baiyee Z M, et al. Nonstoichiometric oxides as low-cost and highly-efficient oxygen reduction/evolution catalysts for low-temperature electrochemical devices [J]. Chem. Rev., 2015, 115: 9869-9921.

[2] Jing Y, Almassi S, Mehraeen S, et al. The roles of oxygen vacancies, electrolyte composition, lattice structure, and doping density on the electrochemical reactivity of Magnéli phase TiO_2 anodes [J]. J. Mater. Chem. A, 2018, 6: 23828-23839.

[3] Geng Z G, Kong X D, Chen W W, et al. Oxygen vacancies in ZnO nanosheets enhance CO_2 electrochemical reduction to CO [J]. Angew. Chem., Int. Ed., 2018, 57: 6054-6059.

[4] Nowotny J. Titanium dioxide-based semiconductors for solar-driven environmentally friendly applications: impact of point defects on performance [J]. Energy Environ. Sci., 2008, 1: 565-572.

[5] Qiu X Q, Miyauchi M, Yu H G, et al. Visible-light-driven Cu(II)-$(Sr_{1-y}Na_y)$-$(Ti_{1-x}Mo_x)O_3$ photocatalysts based on conduction band control and surface ion modification [J]. J. Am. Chem. Soc., 2010, 132: 15259-15267.

[6] Liu M, Qiu X Q, Miyauchi M, et al. Cu(II) oxide amorphous nanoclusters grafted Ti^{3+} self-doped TiO_2: an efficient visible light photocatalyst [J]. Chem. Mater., 2011, 23: 5282-5286.

[7] Liu M, Qiu X Q, Miyauchi M, et al. Energy-level matching of Fe(III) ions grafted at surface and doped in bulk for efficient visible-light photocatalysts [J]. J. Am. Chem. Soc., 2013, 135: 10064-10072.

[8] Xie J F, Zhang H, Li S, et al. Defect-rich MoS_2 ultrathin nanosheets with additional active edge sites for enhanced electrocatalytic hydrogen evolution [J]. Adv. Mater., 2013, 25: 5807-5813.

[9] Radjenovic J, Sedlak D L. Challenges and opportunities for electrochemical processes as next-generation technologies for the treatment of contaminated water [J]. Environ. Sci. Technol., 2015, 49: 11292-11302.

[10] Panizza M, Cerisola G. Direct and mediated anodic oxidation of organic pollutants [J]. Chem. Rev., 2009, 109: 6541-6569.

[11] Martínez-Huitle C A, Ferro S. Electrochemical oxidation of organic pollutants for the wastewater treatment: direct and indirect processes[J]. Chem. Soc. Rev., 2006, 35: 1324-1340.

[12] Yang Y, Hoffmann M R. Synthesis and stabilization of blue-black TiO_2 nanotube arrays for electrochemical oxidant generation and wastewater treatment[J]. Environ. Sci. Technol., 2016, 50: 11888-11894.

[13] Chaplin B P. The prospect of electrochemical technologies advancing worldwide water treatment[J]. Acc. Chem. Res., 2019, 52: 596-604.

[14] Loeb S K, Alvarez P J J, Brame J A, et al. The technology horizon for photocatalytic water treatment: sunrise or sunset? [J]. Environ. Sci. Technol., 2019, 53: 2937-2947.

[15] Chen X, Liu L, Yu P Y, et al. Increasing solar absorption for photocatalysis with black hydrogenated titanium dioxide nanocrystals[J]. Science, 2011, 331: 746-750.

[16] Zuo F, Bozhilov K, Dillon R J, et al. Active facets on titanium(Ⅲ)-doped TiO_2: an effective strategy to improve the visible-light photocatalytic activity[J]. Angew. Chem., Int. Ed., 2012, 51: 6223-6226.

[17] Liu G, Yang H G, Wang X W, et al. Enhanced photoactivity of oxygen-deficient anatase TiO_2 sheets with dominant {001} facets[J]. J. Phys. Chem. C, 2009, 113: 21784-21788.

[18] Ji Y F, Guo W, Chen H H, et al. Surface Ti^{3+}/Ti^{4+} redox shuttle enhancing photocatalytic H_2 production in ultrathin TiO_2 nanosheets/CdSe quantum dots[J]. J. Phys. Chem. C, 2015, 119: 27053-27059.

[19] Liu C, Zhang A Y, Si Y, et al. Photochemical anti-fouling approach for electrochemical pollutant degradation on facet-tailored TiO_2 single crystals [J]. Environ. Sci. Technol., 2017, 51: 11326-11335.

[20] Zhao X, Zhu Y F. Synergetic degradation of rhodamine B at a porous $ZnWO_4$ film electrode by combined electro-oxidation and photocatalysis[J]. Environ. Sci. Technol., 2006, 40: 3367-3372.

[21] Zhao X, Qu J H, Liu H J, et al. Photoelectrocatalytic degradation of triazine-containing azo dyes at γ-Bi_2MoO_6 film electrode under visible light irradiation(λ>420 nm)[J]. Environ. Sci. Technol., 2007, 41: 6802-6807.

[22] Zhao X, Xu T G, Yao W Q, et al. Photoelectrocatalytic degradation of 4-chlorophenol at Bi_2WO_6 nanoflake film electrode under visible light

irradiation[J]. Appl. Catal., B, 2007, 72: 92-97.

[23] Hu L S, Fong C C, Zhang X M, et al. Au nanoparticles decorated TiO_2 nanotube arrays as a recyclable sensor for photo-enhanced electrochemical detection of bisphenol A[J]. Environ. Sci. Technol., 2016, 50: 4430-4438.

[24] Zhang C L, Xu J Q, Li Y T, et al. Photocatalysis-induced renewable field-effect transistor for protein detection [J]. Anal. Chem., 2016, 88: 4048-4054.

[25] Xu J Q, Liu Y L, Wang Q, et al. Photocatalytically renewable micro-electrochemical sensor for real-time monitoring of cells[J]. Angew. Chem., Int. Ed., 2015, 54: 14402-14406.

[26] Xu J Q, Duo H H, Zhang Y G, et al. Photochemical synthesis of shape-controlled nanostructured gold on zinc oxide nanorods as photocatalytically renewable sensors[J]. Anal. Chem., 2016, 88: 3789-3795.

[27] Hu L S, Huo K F, Chen R S, et al. Recyclable and high-sensitivity electrochemical biosensing platform composed of carbon-doped TiO_2 nanotube arrays[J]. Anal. Chem., 2011, 83: 8138-8144.

[28] Li Y H, Liu P F, Pan L F, et al. Local atomic structure modulations activate metal oxide as electrocatalyst for hydrogen evolution in acidic water [J]. Nat. Commun., 2015, 6: 8064.

[29] Swaminathan J, Subbiah R, Singaram V. Defect-rich metallic titania ($TiO_{1.23}$)-an efficient hydrogen evolution catalyst for electrochemical water splitting[J]. ACS Catal., 2016, 6: 2222-2229.

[30] Swaminathan J, Ravichandran S. Insights into the electrocatalytic behavior of defect-centered reduced titania($TiO_{1.23}$)[J]. J. Phys. Chem. C, 2018, 122: 1670-1680.

[31] Feng H, Xu Z, Ren L, Liu C, et al. Activating titania for efficient electrocatalysis by vacancy engineering [J]. ACS Catal., 2018, 8: 4288-4293.

[32] Pelegrini R T, Freire R S, Duran N, et al. Photoassisted electrochemical degradation of organic pollutants on a DSA type oxide electrode: process test for a phenol synthetic solution and its application for the e1 bleach kraft mill effluent[J]. Environ. Sci. Technol., 2001, 35: 2849-2853.

[33] Pelegrini R, Peralta-Zamora P Andrade A R, Reyes J, et al. Electrochemically assisted photocatalytic degradation of reactive dyes[J].

Appl. Catal., B, 1999, 22: 83-90.

[34] Malpass G R P, Miwa D W, Miwa A C P, et al. Photo-assisted electrochemical oxidation of atrazine on a commercial Ti/Ru$_{0.3}$Ti$_{0.7}$O$_2$ DSA electrode[J]. Environ. Sci. Technol., 2007, 41: 7120-7125.

[35] Asmussen R M, Tian M, Chen A C. A new approach to wastewater remediation based on bifunctional electrodes[J]. Environ. Sci. Technol., 2009, 43: 5100-5105.

[36] Qu J H, Zhao X. Design of BDD-TiO$_2$ hybrid electrode with P-N Function for photoelectroatalytic degradation of organic contaminants[J]. Environ. Sci. Technol., 2008, 42: 4934-4939.

[37] Li G T, Qu J H, Zhang X W, et al. Electrochemically assisted photocatalytic degradation of acid orange 7 with β-PbO$_2$ electrodes modified by TiO$_2$[J]. Wat. Res., 2006, 40: 213-220.

[38] Chai S N, Zhao G H, Zhang Y N, et al. Selective photoelectrocatalytic degradation of recalcitrant contaminant driven by an N-P heterojunction nanoelectrode with molecular recognition ability [J]. Environ. Sci. Technol., 2012, 46: 10182-10190.

[39] Chai S N, Zhao G H, Li P Q, et al. Novel sieve-like SnO$_2$/TiO$_2$ nanotubes with integrated photoelectrocatalysis: fabrication and application for efficient toxicity elimination of nitrophenol wastewater[J]. J. Phys. Chem. C, 2011, 115: 18261-18269.

[40] Wang P F, Cao M H, Ao Y H, et al. Investigation on Ce-doped TiO$_2$-coated BDD composite electrode with high photoelectrocatalytic activity under visible light irradiation[J]. Electrochem. Commun., 2011, 13: 1423-1426.

[41] Catanho M, Malpass G R P, Motheo A J. Photoelectrochemical treatment of the dye reactive red 198 using DSA$^@$ electrodes[J]. Appl. Catal., B, 2006, 62: 193-200.

[42] Pinhedoa L, Pelegrinib R, Bertazzolib R, et al. Photoelectrochemical degradation of humic Acid on a (TiO$_2$)$_{0.7}$(RuO$_2$)$_{0.3}$ dimensionally stable anode[J]. Appl. Catal., B, 2005, 57: 75-81.

[43] Zhang A Y, Long L L, Liu C, et al. Electrochemical degradation of refractory pollutants using TiO$_2$ single crystals exposed by high-energy {001} facets[J]. Wat. Res., 2014, 66: 273-282.

[44] Liu S W, Yu J G, Jaroniec M. Anatase TiO$_2$ with dominant high-energy

{001} facets: synthesis, properties and applications[J]. Chem. Mater., 2011, 23: 4085-4093.

[45] Liu G, Yang H G, Pan J, et al. Titanium dioxide crystals with tailored facets[J]. Chem. Rev., 2014, 114: 9559-9612.

[46] Wu M F, Jin Y N, Zhao G H, et al. Electrosorption-promoted photodegradation of opaque wastewater on a novel TiO_2/Carbon aerogel electrode[J]. Environ. Sci. Technol., 2010, 44: 1780-1785.

[47] Liu L, Chen X B. Titanium dioxide nanomaterials: self-structural modifications[J]. Chem. Rev., 2014, 114: 9890-9918.

[48] Chen X B, Liu L, Huang F Q. Black titanium dioxide (TiO_2) nanomaterials[J]. Chem. Soc. Rev., 2015, 44: 1861-1885.

[49] Pan X Y, Yang M Q, Fu X Z, et al. Defective TiO_2 with oxygen vacancies: synthesis, properties and photocatalytic applications[J]. Nanoscale, 2013, 5: 3601-3614.

[50] Su J, Zou X X, Chen J S. Self-modification of titanium dioxide materials by Ti^{3+} and/or oxygen vacancies: new insights into defect chemistry of metal oxides[J]. RSC Adv., 2014, 4: 13979-13988.

[51] Pei D N, Gong L, Zhang A Y, et al. Defective titanium dioxide single crystals exposed by high-energy {001} facets for efficient oxygen reduction[J]. Nat. Commun., 2015, 6: 8696.

[52] Ou G, Xu Y S, Wen B, et al. Tuning defects in oxides at room temperature by lithium reduction[J]. Nat. Commun., 2018, 9: 1302.

[53] Liu C, Zhang A Y, Pei D N, et al. Efficient electrochemical reduction of nitrobenzene by defect-engineered TiO_{2-x} single crystals[J]. Environ. Sci. Technol., 2016, 50: 5234-5242.

[54] Koo M S, Cho K, Yoon J, et al. Photoelectrochemical degradation of organic compounds coupled with molecular hydrogen generation using electrochromic TiO_2 nanotube arrays[J]. Environ. Sci. Technol., 2017, 51: 6590-6598.

[55] Jing Y, Chaplin B P. Mechanistic study of the validity of using hydroxyl radical probes to characterize electrochemical advanced oxidation processes[J]. Environ. Sci. Technol., 2017, 51: 2355-2365.

[56] Jeon T H, Koo M S, Kim H, et al. Dual-functional photocatalytic and photoelectrocatalytic systems for energy-and resource-recovering water treatment[J]. ACS Catal., 2018, 8: 11542-11563.

第 5 章

TiO_{2-x} 单晶的电催化氧还原新功能

第 5 章

TiO_{2-x} 单晶的惰性化学还原新方法

5.1 概述

当前,能源需求日益增长,促使能源转换和存储体系的研究工作广泛开展,以期获得高效率、低成本、环境友好的替代物。氧还原反应(ORR)在金属-空气电池和聚合物膜电解质燃料电池中发挥着关键作用,也是环境催化及能源领域中的重要过程[1-2]。传统的氧还原反应通常需要使用大量 Pt 基催化剂[3-4]。由于 Pt 的成本很高且活性容易衰减,研究人员逐渐开发出性能适宜的非贵金属氧化物(如 Fe_3O_4 和 Co_3O_4 等),作为 Pt 的替代催化剂[5-11]。如之前章节所述,同为非金属氧化物的 TiO_2 具有良好的光、电催化性能,那么是否可以将其引入能源转换与存储体系、开发新功能呢?虽然这一思路极具吸引力,然而,迄今为止仅有的几项关于 TiO_2 用于氧还原反应的研究报道,整体都表现出较低的活性[12-19]。

进一步的研究发现,非化学计量比的还原态 TiO_2 通过形成缺陷结构,即氧空位和 Ti^{3+},能够大幅地降低半导体的禁带宽度(通常小于 2.0 eV),从而增加给体密度、导电率和整体催化性能[20-24]。另外,研究人员证明,Mn^{3+} 具有单个 e_g 占有率,增强了把氧还原反应中间产物稳定在催化剂表面的能力,是氧还原反应的活性价态[25-29]。MnO_2 催化氧还原反应的推动力为 MnO_2 在 O_2 分解中释放电子产生的表面 Mn^{3+} 的化学氧化,因而其催化活性很大程度上依赖于这些物种的电化学氧化还原活性[30-33]。其他 Mn 基氧化物的氧还原反应高活性也被认为是非化学计量比状态和存在表面 Mn^{3+} 的结果[29]。需要特别指出的是,有着独特混合价态的 Mn_3O_4 也被用作活性催化剂,因为与其共存的 $Mn^{2+}/Mn^{3+}/Mn^{4+}$ 促进了缺陷的形成,即空位、电子和空穴,从而决定了微观电子分布[25]。

TiO_2 的表面原子态、理化性质和催化活性也在很大程度上取决于形状和晶面[34]。一方面,与无序堆积的多晶 TiO_2 相比,结晶度高且形貌均一的单晶 TiO_2 通常具有更高的电导率和更低的电子转移内阻,其连续有序的内部晶体结构更有利于反应光载流子的分离转移及整体催化性能的提升。另一方面,与热力学稳定的{101}晶面(低表面自由能 0.44 J·m^{-2})相比,高能{001}晶面(0.90 J·m^{-2})所特有的电子结构几何构型及表面功能基团,会对其稳定性、吸附性质和催化活性产生很大影响。这些特点都促进了{001}-TiO_2 在各类催化反应过程中的有效应用。

由于 TiO_2 与 MnO_2 具有较高的结构相似性，如果合成含有导电性 Ti^{3+}/氧空位的还原态 TiO_{2-x} 单晶，将可以充分探索其作为氧还原反应催化剂的可行性。在本章工作中，通过简便的溶剂热和 H_2 煅烧还原过程合成了晶面设计的氧空位自掺杂的缺陷态 TiO_{2-x} 单晶，结合电化学测试和密度泛函理论计算，系统性地研究了 TiO_{2-x} 单晶作为氧还原反应催化剂的效能；通过氧还原活性及稳定性表征、抗甲醇氧化性测试和缺陷中心反应机理分析，阐释了 TiO_{2-x} 单晶结构与氧还原反应活性之间的构效关系。

5.2 TiO_{2-x} 单晶的合成新方法与氧还原体系设计

5.2.1 TiO_{2-x} 单晶电极的制备与材料学表征

高能{001}晶面还原态 TiO_{2-x} SCs 通过改进的溶剂热方法合成，即在室温条件下，2.0 mL $TiCl_4$ 边搅拌边缓慢加入 40 mL 无水乙醇中，形成透明溶液；然后加入 Ti/Zn 摩尔比为 4∶1 的 Zn 粉末搅拌超过 1.0 h，溶液慢慢变蓝；接着立即转移到 50 mL 聚四氟乙烯内衬的不锈钢高压釜中，在 200 ℃下反应 24 h；结束后冷却到室温，高速离心收集沉淀物，并用乙醇和水反复清洗，再在 60 ℃下真空干燥过夜，最后研磨得到蓝色 Ti^{3+} 自掺杂的 TiO_{2-x} SCs。将商业的 P25（即{101}-TiO_2 PCs）和 TiO_{2-x} SCs 在富氧氛围中 600 ℃煅烧 2.0 h，作为对照。

通过高分辨透射电子显微镜（HRTEM）（JEM-2100，JEOL Co.，Japan）表征形貌结构，X 射线衍射仪（XRD）（XPert，PANalytical BV Co.，Netherlands）分析晶体结构，漫反射光谱（DRS）由紫外-可见分光光度计（UV 2550，Shimadzu Co.，Japan）测量，X 射线光电子能谱仪（XPS，PHI 5600，PerkinElmer Co.，

USA)和 Raman 光谱仪(LABRAM-HR,Horiba JY Co.,Japan)用于表征化学组成,电子自旋共振仪(ESR,JES-FA200,JEOL Co.,Japan)用于提供 Ti 和 O 原子的电子态结构信息,比表面积由 Builder 4200 设备(Tristar Ⅱ 3020M,Micromeritics Co.,USA)通过 BET 方法在液 N_2 温度下测定,FTIR 光谱由光谱仪(Magna-IR 750,Thermo Nicolet Co.,USA)以 KBr 为基底测量。

5.2.2
电催化氧还原测试与计算

所有电化学测量在定制的三电极体系中用电化学工作站(CHI 760D,CH Ins.)实现:工作电极为纯的或改性 TiO_2 沉积的玻碳电极(GCE),对电极为 Pt 丝,参比电极为 Ag/AgCl/饱和 KCl。EIS 在 $10\ g \cdot L^{-1}$ NaCl 溶液中测定,参数如下:交流电压振幅为 5.0 mV,频率范围为 $10^5 \sim 10^{-2}$ Hz。Mott-Schottky 在 $0.1\ mol \cdot L^{-1}$ Na_2SO_4 溶液中通过阻抗测定,参数如下:固有频率为 1000 Hz,电压范围为 0.0~1.0 V/SCE。HER 和 OER 分别在 $0.5\ mol \cdot L^{-1}$ H_2SO_4 和 $0.1\ mol \cdot L^{-1}$ KOH 溶液中通过 LSV 测定,扫速为 $50\ mV \cdot s^{-1}$。

电化学 ORR 的 CV 和 LSV 测试在玻碳旋转圆盘电极(RDE,直径为 5.0 mm,几何表面积为 0.196 cm^2,Pine Res. Ins. Inc.)上进行。2.0 mg TiO_2 样品分散在 2 mL 异丙醇溶剂中,超声 30 min 后,滴加 20 μL 均匀悬浮液,自然风干。TiO_2 纳米粒子通过全氟磺酸(Nafion)溶液(5 μL,0.05 wt%)经过反复滴加直到得到均匀表面,在 RDE 上负载量约 $0.10\ mg \cdot cm^{-2}$[5,9]。电化学 ORR 测试分别在室温下、N_2 或 O_2 饱和(曝气 30 min)的 $0.1\ mol \cdot L^{-1}$ KOH 电解液中测定,电压范围为 -1.3~0.1 V/SCE,扫速为 $10\ mV \cdot s^{-1}$。LSV 需要根据 Pine 模式在不同电极速度下(400~1600 $r \cdot min^{-1}$)测定,而 CV 不需要旋转。在采集电化学数据前,工作电极至少进行 15 次循环,并且在采集时,需要持续通 N_2 或 O_2 来维持电解液中的饱和状态。根据 Koutechy-Levich 模式中的 i^{-1}-$w^{-1/2}$ 线性关系,可以计算 ORR 动力学参数:

$$\frac{1}{i} = \frac{1}{i_k} + \frac{1}{Bw^{1/2}} \tag{5.1}$$

其中,i 为测量的电流,i_k 为动力学电流,w 为电极旋转速度。

Levich 斜率 B 的理论值由如下方程得到:

$$B = 0.62nFC(O_2)\,C(O_2)^{2/3}\,\nu^{-1/6} \tag{5.2}$$

其中，n 为 ORR 的电子转移数，F 为 Faradic 常数（96485 C·mol^{-1}），$C(O_2)$ 为 0.1 mol·L^{-1} KOH 溶液中的 O_2 饱和浓度（1.20×10^{-6} mol·cm^{-3}），$D(O_2)$ 为溶液中 O_2 的扩散系数（1.73×10^{-5} cm^2·s^{-1}），ν 为溶液的运动黏度。

在玻碳盘环电极（RRDE，盘电极的直径为 5.0 mm，几何表面积为 0.196 cm^2，Pine Res. Ins. Inc.）上的电化学 ORR 测试与在 RDE 上方法相同：盘电极电势以 10 mV·s^{-1} 的扫速从负扫到正，环电势恒定在 0.3 V/SCE。$HO_2^-\%$ 和 n 由如下方程计算得到：

$$HO_2^-\% = \frac{200 I_r/N}{I_d + I_r/N} \quad (5.3)$$

$$n = \frac{4 I_d}{I_d + I_r/N} \quad (5.4)$$

其中，I_d 为盘电流；I_r 为环电流；N 为 Pt 环的电流收集系数，根据 $K_3Fe[CN]_6$ 还原测得为 0.40。

5.2.3

电催化氧还原过程的 DFT 理论计算

在 CASTEP 程序中，对电子-离子相互作用采用超软赝势，对超晶胞路径的交换关联泛函采用 Predew-Burke-Ernzerhof 交换的广义梯度近似[35-36]。收敛标准设为快速，每个晶胞的能量允许偏差为 10^{-3} eV。用 CASTEP 的赝势生成器构造范数不变赝势。整个过程基准平面波的能量截断为 750 eV。根据不同 k 点数得到 Brillouin 区集合，而 k 由 Monkhorst-pack 算法生成的单位晶胞尺寸和形状决定。采用有限位移方法计算出声子色散和态密度。运用线性同步传输/二次同步传输方法寻找过渡态结构，并获得 ORR 的活化能 E_a[37]。

吸附能 ΔE_{ads} 用于估算分子-表面相互作用的强度，从而确定能量上最稳定的吸附模型。ΔE_{ads} 表示如下：

$$\Delta E_{ads} = \Delta E_{surf} + \Delta E_{react} - \Delta E_{system} \quad (5.5)$$

其中，总能量值 E_{system}、E_{surf}、E_{react} 分别对应模拟超晶胞中单个分子与锐钛矿型 TiO_2 表面、表面之间，及其与真空中单个孤立分子相互作用。

每个基元反应的 ΔG 变化如下：

$$\Delta G = \Delta E + \Delta ZPVE - T\Delta S \quad (5.6)$$

其中，ΔE 为总能量变化，直接从计算结果得到，$\Delta ZPVE$ 为零点能量变化，T 为热力学温度(298.15 K)，ΔS 为熵变。

表面模型中，为避免厚片与其周期图像之间不必要的相互作用，沿着表面嵌入了一段 15×10^{-10} m 的真空区域。{001}-TiO_2 表面只有通过两个 O 原子桥接的五配位的 Ti 原子，但{101}-TiO_2 表面有通过两个和三个 O 原子桥接的五配位和六配位的 Ti 原子，呈锯齿状排布。{001}-TiO_{2-x} 由{001}-TiO_2 模型基础加入亚表面氧空位而获得。

5.3 TiO_{2-x} 单晶的电催化氧还原性能与机理分析

5.3.1 TiO_{2-x} 单晶的电催化氧还原反应活性

两种 TiO_2 样品(缺陷态与非缺陷态)的形貌基本上都是削顶八面体，粒径约为 20 nm，几乎没有随机堆积的一维纳米棒(图 5.1(a、b))。对缺陷态{001}-TiO_{2-x} SCs，0.35 nm 和 0.47 nm 晶格条纹间距分别对应{101}平面和{002}平面(图 5.1(b))，这表明了削顶后的顶/底暴露面受{001}晶面约束。而且，在内嵌的快速傅里叶变换图中标出夹角为 68.3°，这与{101}平面和{001}平面夹角的理论计算值一致。对非缺陷态{001}-TiO_2 SCs，清晰的晶面间晶格间距和晶面夹角与前者相同(图 5.1(d))。这些结果表明，缺陷态{001}-TiO_{2-x} SCs 在富 O_2 氛围中高温煅烧后，仍保留了形貌、结构和暴露晶面(图 5.1)。

在给定的无 F 醇解路径中，{001}-TiO_{2-x} SCs 通过 $TiCl_4$ 在纯乙醇中受控温和水解形成[38]。可能的反应过程是：乙醇与溶解氧热反应释放出水分子，再通过聚合、凝结(反应(5.7.1))，与 Ti^{4+} 阳离子结合形成钛氯氧化物或羟基氯化物(反应(5.7.2))。随后，这些中间产物逐渐生成 TiO_2(反应(5.8))，在这一步

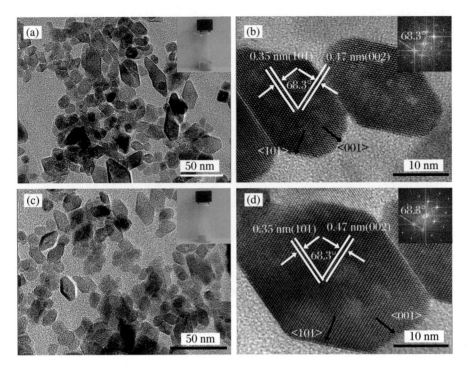

图 5.1　TiO$_2$ 催化剂的形貌与结构：{001}-TiO$_{2-x}$ SCs 和 {001}-TiO$_2$ SCs 的 TEM 图 (a、c) 和 HRTEM 图 (b、d，内嵌为相应的快速傅里叶变换图)

骤中，Cl$^-$ 离子极有可能充当功能基团封端剂与形状控制剂的角色，来有效降低 {001} 晶面和 {101} 晶面的表面能差异[34]。

$$2CH_3CH_2OH + O_2 \rightarrow 2CH_3CHO + 2H_2O \quad (5.7.1)$$

$$TiCl_4 + 2H_2O \rightarrow [Ti(OH)_xCl_n]^{4-(n+x)} \quad (5.7.2)$$

$$\text{or } [TiO_yCl_n]^{4-(n+2y)} \rightarrow TiO_2 + 4HCl \quad (5.8)$$

{101}-TiO$_2$ PCs，TiO$_2$ SCs 和缺陷态 TiO$_{2-x}$ SCs 的 XRD 图的衍射峰均高度符合锐钛矿型 TiO$_2$ (JCPDS No. 21-1272)，而面设计的后两者样品衍射峰增宽可能由相对较小的晶体尺寸引起。另外，缺陷态 TiO$_{2-x}$ SCs 样品有着 Ⅳ 型 N$_2$ 吸附-脱附等温线（计算 BET 比表面积为 47.0 m^2·g^{-1}）和介孔结构磁滞回特征（约 15.0 nm 的窄孔径分布）。这些介孔形成于纳米颗粒聚集成纳米片的过程中。

与商业的 F 调控水热/溶剂热路径相比，该方法合成的 TiO$_2$ SCs 有着更小的粒子尺寸和更大的表面积。尽管高能 {001} 晶面的暴露比例相对较低，但这仍可能为其催化应用提供更多的反应位点。并且，引入 Zn 调控醇热路径合成的 TiO$_{2-x}$ SCs 非常稳定，即使在空气中放置超过一年依然没有观察到明显的颜色变化，可以媲美于其他方法合成的 TiO$_{2-x}$ 样品[20]。金属 Zn 不仅作为形成 Ti^{3+}

的功能还原剂,而且进一步稳定了晶体表面和亚表面新生的氧空位和 Ti^{3+},通过减弱表面 O 键增长了表面氧空位数目,从而增强了缺陷态 TiO_{2-x} 多面体的催化活性[38]。

图 5.2 表明,在 RDE 和 RRDE 上的氧还原反应活性关系为:{001}-TiO_{2-x} SCs>{001}-TiO_2 SCs>{101}-TiO_2 PCs。起始电势在{101}-TiO_2 PCs 上低于 -1.0 V,在{001}-TiO_2 SCs 上正移到约 -0.4 V,在{001}-TiO_{2-x} SCs 上因为氧空位和 Ti^{3+} 自掺杂形成还原态晶体结构,进一步正移到 -0.3 V(图 5.2(a))。同时,阴极电流在面和缺陷设计的 TiO_2 样品上(表 5.1),在给定电势 -0.40 V/SCE 下,测量的电流密度从在{101}-TiO_2 PCs 上的 1.88 $\mu A \cdot cm^{-2}$,增长到在{001}-TiO_2 SCs 上的 39.55 $\mu A \cdot cm^{-2}$,进一步增长到在{001}-TiO_{2-x} SCs 上的 180.06 $\mu A \cdot cm^{-2}$(图 5.2(a))。

三种 TiO_2 样品的 BET 表面积没有显著的区别(对{001}-TiO_{2-x} SCs、{001}-TiO_2 SCs 和{101}-TiO_2 PCs 分别为 47.0 $m^2 \cdot g^{-1}$、43.2 $m^2 \cdot g^{-1}$ 和 42.3 $m^2 \cdot g^{-1}$),它们的聚集模式也相同,表明它们的电化学活性是相当的。因此,观察到的{001}-TiO_{2-x} SCs 优于{001}-TiO_2 SCs 和{101}-TiO_2 PCs 的氧还原反应活性主要归功于其表面缺陷和{001}晶面暴露(图 5.2),而不是其他原因。另外,{001}-TiO_{2-x} SCs 与{001}-TiO_2 SCs 有相同的化学组成、晶体形貌与结构(图 5.1),使得它们具有相当的电化学氧还原反应活性。这表明了观察到的这二者间的氧还原反应活性差别主要源自晶体缺陷、氧空位和还原态 Ti^{3+},而不是其他原因。再者,非整数比的{001}-TiO_{2-x} SCs 不饱和配位的 Ti 位点增强了载流子密度,有利于 O_2 沿着导电晶体通道的分子扩散。因为 Zn^{2+} 的离子半径 $(0.88 \times 10^{-10}$ m$)$ 大于 Ti^{4+} $(0.745 \times 10^{-10}$ m$)$[38],引入的兼作还原剂与稳定剂的 Zn,主要以 ZnO 团簇的形式散布在 TiO_2 表面。而 ZnO 作为宽禁带宽度 N 型半导体,电化学活性很低,在本工作中引入很少量(原子比约为 0.80%)的 ZnO 杂质并不会干扰氧还原反应[39-40](图 5.3)。

根据 Koutecky-Levich 模型从不同电势下的直线斜率计算出氧还原反应的电子转移数 n,发现 n 随着超电势增加而持续增加。计算得到缺陷{001}-TiO_{2-x} SCs 在 -0.4 V 和 -1.3 V/SCE 时,电子转移数为 0.44 和 2.41(图 5.2(c)),但{001}-TiO_2 SCs 和{101}-TiO_2 SCs 分别仅有 0.26~2.34 和 0.00~0.59。以上计算结果表明,在 -0.4 V 到 -1.3 V/SCE 条件下,发生在缺陷{001}-TiO_{2-x} SCs 上的阴极氧还原反应主要由高超电势下的两电子路径主导。

通过 RRDE 结果进一步验证上述两电子路径。结论与密度泛函理论计算高度一致。在 RRDE 测试中,电解液为 0.1 mol $\cdot$ L^{-1} KOH,缺陷态{001}-

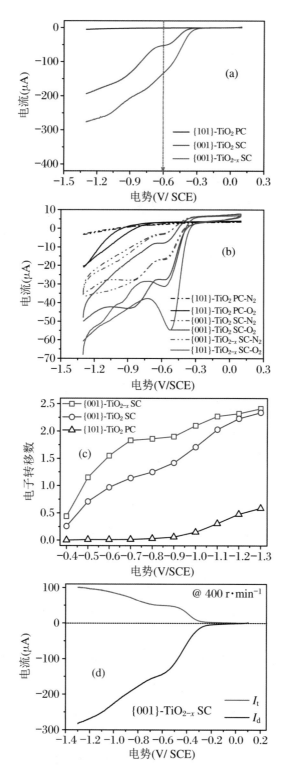

图 5.2 TiO₂ 催化剂的氧还原反应活性：LSV 曲线(a、e)，CV 曲线(b、f)，电子转移数(c)，在 -0.45 V/SCE 条件下、O₂ 饱和的 0.1 mol·L⁻¹ KOH 溶液中、添加 10 vol% 甲醇前后的 I-t 曲线(g)和阴极电流稳定性曲线(h)

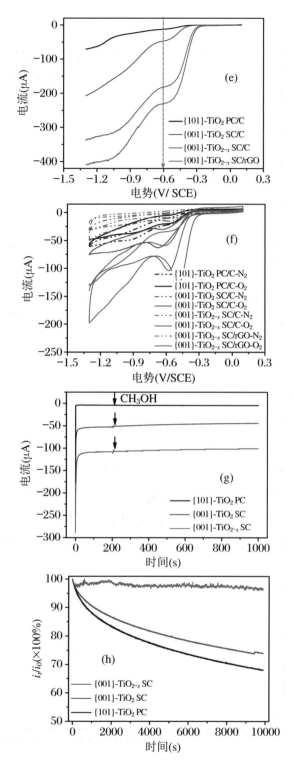

续图 5.2　TiO_2 催化剂的氧还原反应活性:LSV 曲线(a、e),CV 曲线(b、f),电子转移数(c),在 -0.45 V/SCE 条件下、O_2 饱和的 0.1 mol·L^{-1} KOH 溶液中、添加 10 vol%甲醇前后的 I-t 曲线(g)和阴极电流稳定性曲线(h)

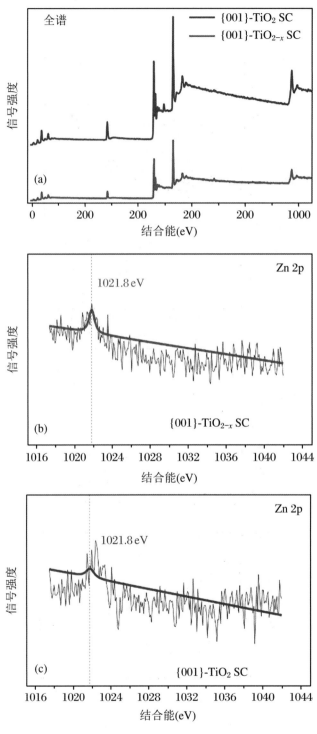

图 5.3　TiO$_2$ SCs 和 TiO$_{2-x}$ SCs 的 XPS 谱图：(a) 总谱；
(b) TiO$_2$ SCs 的 Zn 2p 分谱；(c) TiO$_{2-x}$ SCs 的 Zn 2p 分谱

TiO_{2-x} SCs 修饰的电极在起始电势开始的整个电势范围内产生了显著的环电流 I_r。这表明在该电化学还原过程中 HO_2^- 中间产物的生成反应占主导。计算得到缺陷态 {001}-TiO_{2-x} SCs 在 −0.4 V 和 −1.3 V 时，H_2O_2 产率即 HO_2^- % 分别为 111.0% 和 94.2%，电子转移数为 1.78~2.17。以上结果与 RDE 测试一致，也表明在缺陷态 {001}-TiO_{2-x} SCs 上的氧还原反应主要通过两电子路径发生。

考虑到金属氧化物催化剂的电子导电性低和分散性差，在氧还原测试中通常需要负载在纳米结构的 C 基底上。本书所有的 TiO_2 样品都与作为导电基底的商业 Cabot CNPs 混合。观察到相同的氧还原反应活性趋势：{001}-TiO_{2-x} SCs/C＞{001}-TiO_2 SCs/C＞{101}-TiO_2 PCs/C（表 5.1）。继而发现，通过水热反应将 {001}-TiO_{2-x} SCs 沉积在石墨烯上形成的复合物，其氧还原反应能力进一步增强（图 5.2(e、f)）。石墨烯与金属氧化物的相互作用带来的沿着界面的电荷转移，与复合物界面间距及其 Fermi 能级差相关，可能是增强的氧还原反应活性的主要成因[10]。此外，所有的 LSV 有缓慢的电流增长且没有平台期，同样表明了 O_2 通过两电子路径还原到了 HO_2^-。以上结果清晰地表明了暴露的高能 {001} 晶面和氧空位设计均能为 TiO_2 催化剂增强的氧还原反应活性作贡献。

氧还原反应的稳定性通过在 −0.45 V/SCE 下的计时安培响应来评估，电解液为 O_2 饱和的 0.1 mol·L^{-1} KOH（图 5.2(g、h)）。缺陷态 {001}-TiO_{2-x} SCs 有着极好的稳定性，在 10000 s 后仍能保留 95% 的原始电流，而 {001}-TiO_2 SCs 和 {101}-TiO_2 PCs 都损失了接近 30% 的原始电流。此外，所有的 TiO_2 样品对氧还原反应都有着极高的选择性，在加入甲醇后都没有明显的电流变化。再者，在高度腐蚀的电解液 6.0 mol·L^{-1} KOH 中，{001}-TiO_{2-x} SCs 仍能继续发挥作用，证实了它的高活性、有利动力学和出色的稳定性。观察到的缺陷态 TiO_{2-x} SCs 的出色稳定性主要归功于表面设计的 Ti^{3+}[38]，而 {101}-TiO_2 PCs 的稳定性很差可能是由其锐钛矿-金红石混合相导致的[41-42]。

综上结果可见，TiO_2 的氧还原反应活性可以通过简单的面和缺陷设计策略得到大幅增强。需要注意的是，该工作中合成的高能 {001} 晶面暴露的缺陷态 TiO_{2-x} 单晶的氧还原反应活性依然低于最典型的金属氧化物氧还原电催化剂 Fe_3O_4 和 Mn_3O_4[10]，它的起始电势更负，大约为 −0.30 V，阴极电流密度也更低，大约为 1.0 mA·cm^{-2}（图 5.2(a) 和 (b)）。另外，它的氧还原反应稳定性比商业 Pt/C 基准高得多，计时安培响应几乎没有变化，而 Pt/C 损失了超过 20%[4]。由于 TiO_2 储量丰富、低格价廉、稳定性强且环境相容性好，因此它在工业生产中具有良好的应用前景。

5.3.2
TiO$_{2-x}$ 单晶中 Ti^{3+} 的存在与作用

表面 Ti 氧化态通过 FTIR 光谱、DRS 谱、XPS 能谱、ESR 谱和 Raman 光谱证实(图 5.4)。FTIR 谱中,在 3400 cm^{-1} 和 1632 cm^{-1} 处对应的—OH 伸缩振动特征峰,表明了作为水分子和—OH 反应吸收位点的氧空位和/或还原态 Ti^{3+} 的存在(图 5.4(a))。同时较宽的吸收,表明了—OH 在缺陷态{001}-TiO$_{2-x}$ SCs 中所处的环境更复杂。高分辨 XPS 谱中,Ti 2p 谱在低能区(457.5~455.5 eV)间出现一个明显的肩峰,这可能归因于 TiO$_2$ 晶格中的 Ti^{3+} (图 5.4(c、d))[22]。ESR 谱中在 $g=1.95$ 和 2.00 处的强烈的表面 Ti^{3+} 信号进一步证实了这点[22]。而该信号在 $g=2.02$ 处由于表面氧化消失了(图 5.4(e))。此外,Raman 谱中由改进的几何与表面结构引起的峰正移也证实了还原态 TiO$_{2-x}$ SCs 存在更多的氧空位(图 5.4(f))[23]。

原则上,Ti^{3+} 是由还原条件下 Ti^{4+} 的还原生成的,并且原位 Ti^{3+} 掺杂的 TiO$_2$ 由于晶格中 Ti^{3+} 取代了 Ti^{4+} 出现了氧空位[20]。同时,{001}晶面上每个 Ti 与五个氧原子配位,而{101}晶面可能与五或六个氧原子配位(概率约为 1∶1),所以{001}晶面比{101}晶面包含更多的氧空位[34]。Ti^{3+} 和氧空位形成了一些低于 TiO$_2$ 导带的新能级[22],导致可见光范围内的光谱发生明显变化(图 5.4(b))。而在红外区域的光子激发也表明了形成的 Ti^{3+} 主要源于氧缺陷[24]。另外,氧缺陷态 TiO$_{2-x}$ SCs 与非缺陷态 TiO$_2$ SCs 的 Raman 谱不同,但 X 射线衍射(XRD)图相近,表明了大多数与氧缺陷相关的 Ti^{3+} 主要分布在晶体的表面和亚表面[21]。

以上这些结构缺陷对 TiO$_{2-x}$ 的电催化性质起着关键作用。在还原态 TiO$_{2-x}$ SCs(图 5.4(b))上增强的可见光吸收,表明了 Ti^{3+} 自掺杂引入的新能态大都低于导带底,而这些导电的 Ti^{3+} 正能够降低固有带隙和增强电子传导率(图 5.5(a))。除外,两种 TiO$_2$ 样品(缺陷态与非缺陷态)的 Mott-Schotty 曲线都具有正斜率,是为 N 型半导体,而且 TiO$_{2-x}$ SCs 有更高的电子密度(图 5.5(b))[21]。因此,虽然 TiO$_{2-x}$ SCs 不是直接带隙半导体(图 5.4(b)),在自掺杂过程中发生的电子结构改进使得其电子密度变高。而且,TiO$_{2-x}$ SCs 的 EIS 的第二个半圆(与快速反应动力学相关,在可及电化学表面增加相当大)显示了增强的电化学导电率、降低的电子转移电阻和增强的界面电容,这都是高效氧还原催化剂的必要条件。

同时降低 HER 和 OER 的电化学超电势也有利于其应用。本工作中，缺陷态{001}-TiO$_{2-x}$ SCs 上的 HER 和 OER(图 5.5(c、d))的超电势大幅降低，电流也大幅增强。例如，{101}-TiO$_2$ PCs 在 -0.3 V/SCE 处的阴极电流和在 1.2 V/SCE 处的阳极电流分别为 -1.9 μA 和 12.3 μA，{001}-TiO$_2$ SCs 分别为 -7.5 μA 和 20.2 μA，{001}-TiO$_{2-x}$ SCs 分别为 -25.7 μA 和 185.6 μA。催化活性为：{001}-TiO$_{2-x}$ SCs＞{001}-TiO$_2$ SCs＞{101}-TiO$_2$ PCs。该结果表明，在 TiO$_2$ 晶格均匀地引入氧空位缺陷和暴露高能{001}晶面能够大幅改善其 HER 和 OER 催化性能。

5.3.3
TiO$_{2-x}$ 单晶的电催化氧还原反应机理

氧还原反应受化学组成、元素价态、晶体结构和表面态影响[5]。表面氧分子的吸附构型及其相应的氧分子键强也有影响。本工作中，缺陷态{001}-TiO$_{2-x}$ SCs 上增强的氧还原活性与单晶结构、暴露的高能{001}晶面和由氧空位缺陷引起的 Ti 混合价密切相关(图 5.6)。首先，连续有序的内部结构和增强的结晶度使得电子传导率增强，从而加速了电子转移，意味着更高的电荷迁移率，并减少了氧还原反应中的电极极化；其次，暴露的高能{001}晶面具有独特的原子、电子和能量结构，在解离吸附和电荷迁移中有面导向性，保证了其氧还原活性[34]；最后，介于其单个 e$_g$ 占有率及其增强的把中间产物稳定在催化剂表面的能力，由设计的氧空位带来的缺陷 Ti^{3+} 活跃价态在氧还原反应中起支配作用[43]。

在 TiO$_2$ 晶体结构中，体或表面丢失的氧原子被一个或两个"自由"电子占据，三个最近的 Ti 原子趋向于远离空位，以便于增强其与其余晶格的键强[20]。因而，与缺陷相关的性质主要包括结构、电子性质、光学性质、解离吸附性质和还原性质[21]。对于氧还原反应，O$_2$ 在 TiO$_2$ 表面的化学吸附和解离激活是控制步骤。已有结果证明这两个过程都只在热力学上有利[22]。一般而言，由于被吸附物和表面的强结合，O$_2$ 在氧空位缺陷上的解离吸附带来更高效的电子转移过程和更大的氧还原反应动力学系数，因而与整体催化性能紧密相关。

在缺陷态{001}-TiO$_{2-x}$ SCs 上，由于其很高的吸附能量，O$_2$ 首先选择性地吸附在能够形成 O—Ti 键的带有过剩负电的氧空位缺陷亚表面(反应(5.9))。然后化学吸附的 O$_2$ 捕捉到氧空位位点上的自由电子，同时生成超氧自由基团

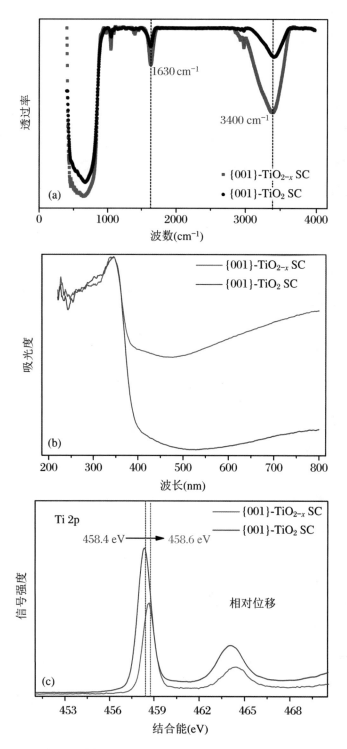

图 5.4 TiO$_{2-x}$ 单晶中的 Ti^{3+} 的存在：{001}-TiO$_{2-x}$ SCs 和 {001}-TiO$_2$ SCs 的 FTIR(a)、DRS(b)、XPS(c、d、g、h)、ESR(e)、Raman(f) 谱图

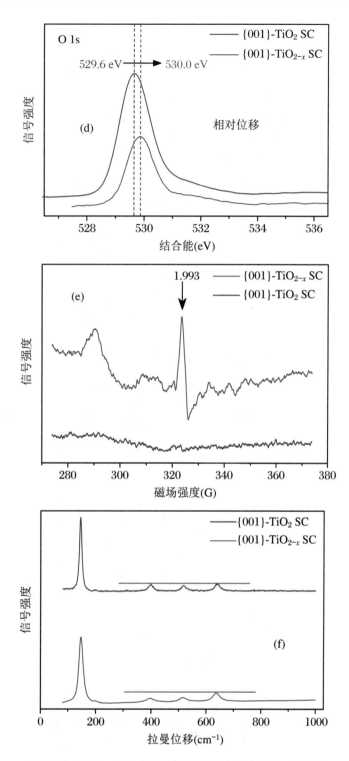

续图 5.4　TiO_{2-x} 单晶中的 Ti^{3+} 的存在：{001}-TiO_{2-x} SCs 和 {001}-TiO_2 SCs 的 FTIR(a)、DRS(b)、XPS(c、d、g、h)、ESR(e)、Raman(f)谱图

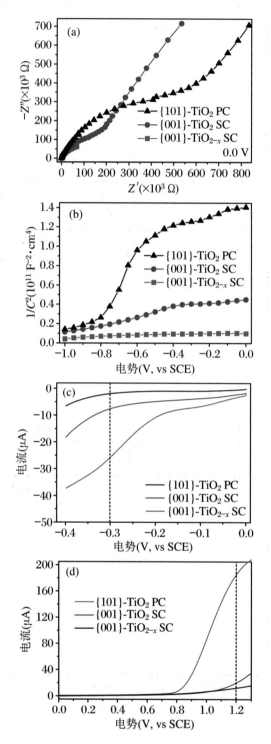

图 5.5 TiO$_{2-x}$ 单晶中 Ti^{3+} 诱导的电子特性和电化学性质：{101}-TiO$_2$ PCs、{001}-TiO$_2$ SCs 和 {001}-TiO$_{2-x}$ SCs 的 EIS(a)、Mott-Schottky(b)、HER(c) 和 OER(d) 曲线

(反应(5.10))。这些自由基团能够有效地促进电荷分离以及通过两电子路径(反应(5.11)、(5.12.1)和(5.12.2))或四电子路径(反应(5.13)、(5.14.1)、(5.14.2)和(5.14.3))的 O_2 还原,紧接着 Ti^{4+} 被还原到 Ti^{3+}(反应(5.15))。在此缺陷中心的氧还原反应中(图5.6),可能发生了 H_2O_2 催化歧化,它成为整个氧还原反应的速度限制步骤。

$$Ti^{3+} + O_2 \rightarrow Ti^{3+} - O_2^* \tag{5.9}$$

$$Ti^{3+} - O_2^* \rightarrow Ti^{4+} + O_2^{-*} \tag{5.10}$$

$$\cdot O_2^{-*} + H_2O + Ti^{3+} \rightarrow HO_2^* + OH^- + Ti^{4+} \tag{5.11}$$

$$\cdot O_2^{-*} + Ti^{3+} \rightarrow \cdot O_2^{2-*} + Ti^{4+} \tag{5.12.1}$$

$$\cdot O_2^{2-*} + H_2O \rightarrow HO_2^* + OH^- \tag{5.12.2}$$

$$HO_2^* + H_2O + 2\,Ti^{3+} \rightarrow 3\,OH^- + 2\,Ti^{4+} \tag{5.13}$$

$$HO_2^* \rightarrow O^* + OH^- \tag{5.14.1}$$

$$O^* + H_2O + Ti^{3+} \rightarrow OH^* + OH^- + Ti^{4+} \tag{5.14.2}$$

$$OH^* + Ti^{3+} \rightarrow H_2O + Ti^{3+} \rightarrow OH^- + Ti^{4+} \tag{5.14.3}$$

$$Ti^{4+} + e^- \rightarrow Ti^{3+} \tag{5.15}$$

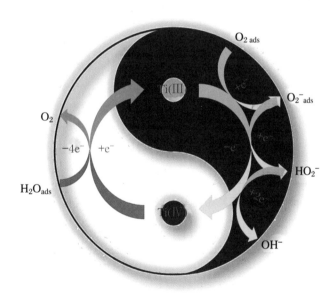

图 5.6　TiO_{2-x} 单晶中的缺陷中心氧还原反应机理

缺陷晶体氧空位上的 Ti^{3+}/Ti^{4+} 混合价态使其从热力学上成为阴极反应的活性位点

密度泛函理论计算进一步证实了氧空位包围的四和五配位 Ti 位点分别在 O_2 吸附和反应活化中起着降低活化能(E_a)和避免催化中毒的作用。一方面,发现 O_2 以最高的吸附能量(2.28 eV)(表5.2)与 {001}-TiO_{2-x} SCs 牢牢结合。另一方面,进一步计算了以最优化的几何吸附构型(图 5.7)为基础的关键中间产物 O_2^{-*}、O_2^{2-*}、HO_2^{-*}、OH^*、O^*($*$表示吸附态)(表 5.1)的自由能(G)和图 5.8(a)中氧还原反应的每一步的自由能变化(ΔG)(表 5.3),并在图 5.8(b)标绘出最有利的能量曲线,确定在{001}-TiO$_{2-x}$ SCs 上。首先,相比于在{001}-TiO$_2$ SCs 上的-2.478 eV 和在{101}-TiO$_2$ PCs 上的 3.945 eV,在{001}-TiO$_{2-x}$ SCs 上吸附 O_2 形成 O_2^{-*} 的 ΔG 最低,为-3.296 eV;其次,相比于在{001}-TiO$_2$ SCs 上的 7.550 eV 和在{101}-TiO$_2$ PCs 上的 10.582 eV,在{001}-TiO$_{2-x}$ SCs 上从生成的吸附 O 中再生 OH$^-$(整个氧还原反应的限速步)的 E_a 最低,为 5.007 eV;最后,相比于在{001}-TiO$_2$ 上的-3.608 eV 和在{101}-TiO$_2$ PCs 上的-2.806 eV,在{001}-TiO$_2$ SCs 上从吸附 O_2 到 4OH$^-$ 的总 ΔG 为-7.741 eV。这些结果不仅证实了{001}-TiO$_2$ SCs 比{101}-TiO$_2$ PCs 热力学和动力学更活跃,而且显示了{001}-TiO$_{2-x}$ SCs 的氧空位有效地增强了其催化活性。该结论与电化学结果一致(图 5.2)。

表 5.1 TiO$_2$ 在 O_2 饱和的 0.1 mol·L^{-1} KOH 溶液中的氧还原反应比活性(SA)比较

催化剂	比活性($\times 10^{-3}$ mA·cm^{-2})		
	-0.35 V	-0.40 V	-0.45 V
{101}-TiO$_2$ PCs	-1.50	-1.88	-2.41
{001}-TiO$_2$ SCs	-12.11	-39.55	-106.65
{001}-TiO$_{2-x}$ SCs	-62.17	-180.06	-349.41
{101}-TiO$_2$ PCs/C	-2.68	-9.43	-24.44
{001}-TiO$_2$ SCs/C	-10.13	-34.89	-95.23
{001}-TiO$_{2-x}$ SCs/C	-180.82	-470.90	-748.66
{001}-TiO$_{2-x}$ SCs/rGO	-269.51	-622.19	-943.19

注:扫速为 10 mV·s^{-1},电极转速为 400~1600 r·min^{-1}。

表 5.2 {001}-TiO$_{2-x}$ 和{001}-TiO$_2$ 上三种不同吸附构型的吸附能(ΔE_{ads})

催化剂	E_{surf}(eV)	E_{react}(eV)	E_{system}(eV)	ΔE_{ads}(eV)
{001}-TiO$_{2-x}$ SCs	-20794.63	-857.85	-21654.76	2.28
{001}-TiO$_2$ SCs	-21226.38	-857.85	-22085.26	1.03

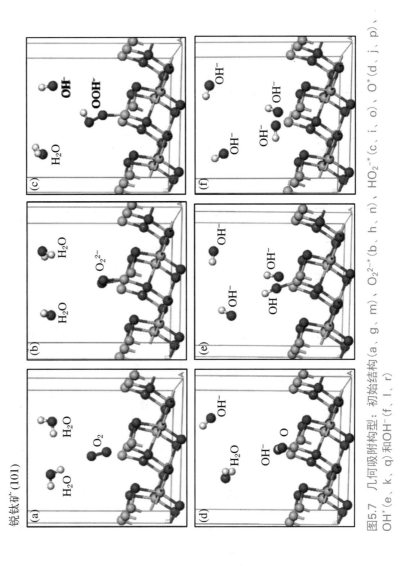

图5.7 几何吸附构型：初始结构(a、g、m)、O_2^{2-*}(b、h、n)、HO_2^{-*}(c、i、o)、O^*(d、j、p)、OH^*(e、k、q)和OH^-(f、l、r)

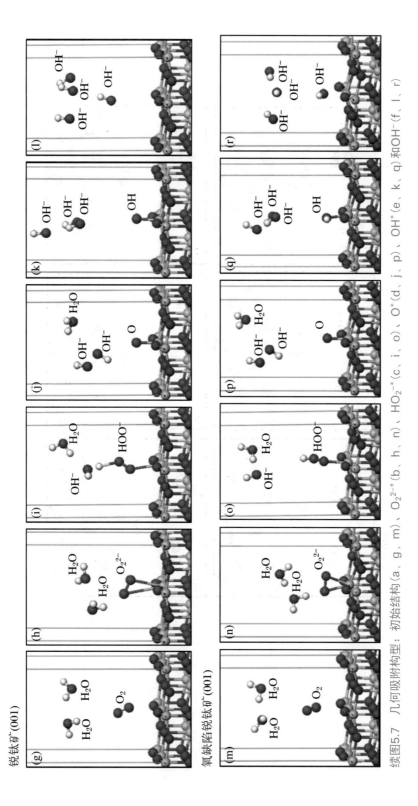

续图5.7 几种吸附构型：初始结构（a、g、m）、O_2^{2-*}（b、h、n）、HO_2^{-*}（c、i、o）、O^*（d、j、p）、OH^*（e、k、q）和OH^-（f、l、r）

表 5.3　TiO_2 的计算热力学性质(ΔG)和氧还原反应能垒(E_a)(298 K/1 atm)

ORR 步骤	{001}-TiO_2 SCs		{001}-TiO_{2-x} SCs		{101}-TiO_2 PCs	
	ΔG	E_a	ΔG	E_a	ΔG	E_a
$O_2^* + 2e^- \rightarrow O_2^{2-*}$	-2.478	—	-3.296	—	3.945	—
$O_2^{2-*} + H_2O \rightarrow HO_2^{-*} + OH^-$	-0.833	-0.091	-0.923	2.889	-3.400	2.950
$HO_2^{-*} \rightarrow O^* + OH^-$	2.648	2.900	1.372	2.030	-3.400	0.800
$O^* + H_2O + e^- \rightarrow OH^* + OH^-$	-2.806	7.550	-4.431	5.007	-2.583	10.582
$OH^* + e^- \rightarrow OH^-$	-0.139	0.430	-0.463	-0.302	1.887	-3.080
总反应	-3.608	—	-7.741	—	-3.551	—

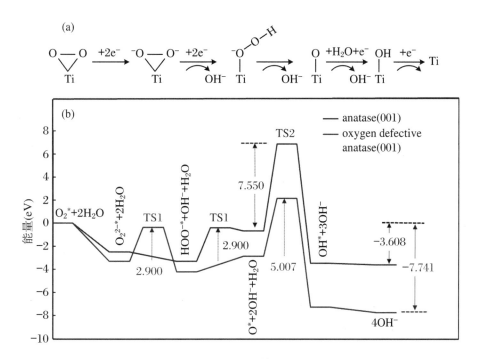

图 5.8　氧还原反应机理和能量图:(a) DFT 计算得到的锐钛矿型 TiO_2 表面的氧还原反应机理;(b) {001}-TiO_2 SCs 和 {001}-TiO_{2-x} SCs 的氧还原反应的能量图

虽然在{001}-TiO_{2-x} SCs 上生成的吸附态 HO_2^- 二次还原有着 5.01 eV 的高 E_a,但其 ΔG 为 -4.43 eV(表 5.3),表明在室温下此步骤能够自发进行,尽管反应速率很低。因此,氧还原反应平衡能够逐渐转向深度氧还原。密度泛函理论计算出的该高能垒的氧还原反应限速步为从生成的 HO_2^- 的二次还原,进一步经过了 RDE 和 RRDE 测试的验证。计算得到,即使电势在很负的 -1.3 V/SCE,缺陷态{001}-TiO_{2-x} SCs 的电子转移数和 HO_2^- %分别约为 2.20% 和 95%(图 5.2(c))。另外,该限速步的几何结构变化很简单,由此后 H_2O 分解所生成的吸附态 OH^* 引起。因此可以确认,两电子机理体系是该步反应的氧还原途径。

缺陷态{001}-TiO$_{2-x}$ SCs 作为缺陷中心氧还原反应最活跃的催化剂,其混合价态也符合之前钙钛矿催化剂的设计原则,特别是在 Mn 基氧化物。而且,氧还原反应涉及表面缺陷 Ti 的氧化和还原,导致这些氧化还原中心的数目和活性都会影响整体催化性能。

本章小结

本章工作证明,通过简捷晶面调控和缺陷设计的 TiO$_2$ 单晶具备出色的电催化氧还原反应活性、稳定性和抗甲醇氧化性。相较非缺陷态 TiO$_2$ 单晶和 TiO$_2$ 多晶,该材料的起始电势分别正移了 0.1 V 和 0.6 V;在 -0.40 V/SCE 条件下,电流密度分别增长了近 4.6 倍和 10 倍。在持续反应 10000 s 后,该材料仍能保留 95% 的原始电流,且在加入甲醇后没有明显的电流变化。单晶结构、高能 {001} 晶面暴露以及 Ti^{3+} 和氧空位的存在,都对 TiO$_2$ 电极活性的提升发挥了重要作用。首先,连续有序的单晶结构和增强的结晶度加速了电子传导,减少了电极极化;其次,高能 {001} 晶面的暴露有利于氧还原反应过程中 O$_2$ 的界面吸附和解离激活,而表面氧空位和还原态 Ti^{3+} 位点的存在,保障了氧还原反应能在缺陷中心温和稳定地进行;最后,与石墨烯的有效复合,提高了分散性和导电性,进一步提升了氧还原反应性能。计算结果表明,氧还原反应的限速步为中间产物 HO$_2^-$ 的二次还原,同时证明了总反应为两电子路径。

参考文献

[1] Adler S B. Factors governing oxygen reduction in solid oxide fuel cell cathodes[J]. Chem. Rev., 2004, 104: 4791-4843.

[2] Zhang J, Zhao Z, Xia Z, et al. A metal-free bifunctional electrocatalyst for oxygen reduction and oxygen evolution reactions[J]. Nat. Nanotechnol., 2015, 10: 444-452.

[3] Cheng F, Shen J, Peng B, et al. Rapid room-temperature synthesis of nanocrystalline spinels as oxygen reduction and evolution electrocatalysts[J].

Nat. Chem., 2011, 3: 79-84.

[4] Li Y, Zhou W, Wang H, et al. An oxygen reduction electrocatalyst based on carbon nanotube-graphene complexes[J]. Nat. Nanotechnol., 2012, 7: 394-400.

[5] Liang Y, Li Y, Wang H, et al. Co_3O_4 nanocrystals on graphene as a synergistic catalyst for oxygen reduction reaction[J]. Nat. Mater., 2011, 10: 780-786.

[6] Suntivich J, Gasteiger H A, Yabuuchi N, et al. Design principles for oxygen-reduction activity on perovskite oxide catalysts for fuel cells and metal-air batteries[J]. Nat. Chem., 2011, 3: 546-550.

[7] Cheng F, Zhang T, Zhang Y, et al. Enhancing electrocatalytic oxygen reduction on MnO_2 with vacancies[J]. Angew. Chem., Int. Ed., 2013, 52: 2474-2477.

[8] Cheng F Y, Chen J. Transition metal vanadium oxides and vanadate materials for lithium batteries[J]. J. Mater. Chem., 2011, 21: 9841-9848.

[9] Liang Y, Wang H, Diao P, et al. Oxygen reduction electrocatalyst based on strongly coupled cobalt oxide nanocrystals and carbon nanotubes[J]. J. Am. Chem. Soc., 2012, 134: 15849-15857.

[10] Wu Z S, Yang S, Sun Y, et al. 3D nitrogen-doped graphene aerogel-supported Fe_3O_4 nanoparticles as efficient electrocatalysts for the oxygen reduction reaction[J]. J. Am. Chem. Soc., 2012, 134: 9082-9085.

[11] Guo S, Zhang S, Wu L, et al. Co/CoO nanoparticles assembled on graphene for electrochemical reduction of oxygen[J]. Angew. Chem., Int. Ed., 2012, 51: 11770-11773.

[12] Liu B, Chen H M, Liu C, et al. Large-scale synthesis of transition-metal-doped TiO_2 nanowires with controllable overpotential[J]. J. Am. Chem. Soc., 2013, 135: 9995-9958.

[13] Choi Y K, Seo S S, Chjo K H, et al. Thin Titanium-dioxide film electrodes prepared by thermal-oxidation[J]. J. Am. Chem. Soc., 1992, 139: 1803-1807.

[14] Kim J H, Ishihara A, Mitsushima S, et al. Catalytic activity of titanium oxide for oxygen reduction reaction as a non-platinum catalyst for PEFC[J]. Electrochim. Acta, 2007, 52: 2492-2497.

[15] Boskovic I, Mentus S V, Pjescic M. Electrochemical behavior of an Ag/

TiO₂ composite surfaces[J]. Electrochim. Acta, 2006, 51: 2793-2799.

[16] Berger T, Monllor-Satoca D, Jankulovska M, et al. The electrochemistry of nanostructured titanium dioxide electrodes[J]. Chemphyschem, 2012, 13: 2824-2875.

[17] Tammeveski K. The Reduction of Oxygen on Pt-TiO₂ coated Ti electrodes in alkaline solution[J]. J. Electrochem. Soc., 1999, 146: 669.

[18] Mentus S V. Oxygen reduction on anodically formed titanium dioxide[J]. Electrochim. Acta, 2004, 50: 27-32.

[19] Wang B. Recent development of non-platinum catalysts for oxygen reduction reaction[J]. J. Power Sources, 2005, 152: 1-15.

[20] Pan X, Yang M Q, Fu X, et al. Defective TiO₂ with oxygen vacancies: synthesis, properties and photocatalytic applications[J]. Nanoscale, 2013, 5: 3601-3614.

[21] Wang G, Ling Y, Li Y. Oxygen-deficient metal oxide nanostructures for photoelectrochemical water oxidation and other applications[J]. Nanoscale, 2012, 4: 6682-6691.

[22] Nowotny J. Titanium dioxide-based semiconductors for solar-driven environmentally friendly applications: impact of point defects on performance[J]. Energy Environ. Sci., 2008, 1: 565-572.

[23] Xing M Y, Fang W Z, Nasir M, et al. Self-doped Ti^{3+}-enhanced TiO₂ nanoparticles with a high-performance photocatalysis[J]. J. Catal., 2013, 297: 236-243.

[24] Liu G, Yang H G, Wang X W, et al. Enhanced photoactivity of oxygen-deficient anatase TiO₂ sheets with dominant {001} facets[J]. J. Phys. Chem. C, 2009, 113: 21784-21788.

[25] Song M K, Cheng S, Chen H, et al. Anomalous pseudocapacitive behavior of a nanostructured, mixed-valent manganese oxide film for electrical energy storage[J]. Nano Lett., 2012, 12: 3483-3490.

[26] Lima F H B, Calegaro M L, Ticianelli E A. Investigations of the catalytic properties of manganese oxides for the oxygen reduction reaction in alkaline media[J]. J. Electroanal. Chem., 2006, 590: 152-160.

[27] Han X P, Zhang T R, Du J, et al. Porous calcium-manganese oxide microspheres for electrocatalytic oxygen reduction with high activity[J]. Chem. Sci., 2013, 4: 368-376.

[28] Gorlin Y, Jaramillo T F. A bifunctional nonprecious metal catalyst for oxygen reduction and water oxidation[J]. J. Am. Chem. Soc., 2010, 132: 13612-13614.

[29] Stoerzinger K A, Risch M, Suntivich J, et al. Oxygen electrocatalysis on {001}-oriented manganese perovskite films: Mn valency and charge transfer at the nanoscale[J]. Energy Environ. Sci., 2013, 6: 1582-1588.

[30] El-Deab M S, Ohsaka T. Manganese oxide nanoparticles electrodeposited on platinum are superior to platinum for oxygen reduction[J]. Angew. Chem., Int. Ed., 2006, 45: 5963-5966.

[31] Xiao W, Wang D L, Lou X W. Shape-controlled synthesis of MnO_2 nanostructures with enhanced electrocatalytic activity for oxygen reduction[J]. J. Phys. Chem. C, 2010, 114: 1694-1700.

[32] Duan J, Zheng Y, Chen S, et al. Mesoporous hybrid material composed of Mn_3O_4 nanoparticles on nitrogen-doped graphene for highly efficient oxygen reduction reaction[J]. Chem. Commun., 2013, 49: 7705-7707.

[33] Tian Z R, Tong W, Wang J Y, et al. Manganese oxide mesoporous structures: mixed-valent semiconducting catalysts[J]. Science, 1997, 276: 926-930.

[34] Liu S, Yu J, Jaroniec M. Anatase TiO_2 with dominant high-energy {001} facets: synthesis, properties, and applications[J]. Chem. Mater., 2011, 23: 4085-4093.

[35] Perdew J P, Burke K, Ernzerhof M. Generalized gradient approximation made simple[J]. Phys. Rev. Lett., 1996, 77: 3865-3868.

[36] Aschauer U, He Y B, Cheng H Z, et al. Influence of subsurface defects on the surface reactivity of TiO_2: water on anatase {101}[J]. J. Phys. Chem. C, 2010, 114: 1278-1284.

[37] Halgren T A, Lipscomb W N. The synchronous-transit method for determining reaction pathways and locating molecular transition states[J]. Chem. Phys. Lett., 1977, 49: 225-232.

[38] Zheng Z, Huang B, Meng X, et al. Metallic zinc-assisted synthesis of Ti^{3+} self-doped TiO_2 with tunable phase composition and visible-light photocatalytic activity[J]. Chem. Commun., 2013, 49: 868-870.

[39] Goux A, Pauporté T, Lincot D. Oxygen reduction reaction on electrodeposited zinc oxide electrodes in KCl solution at 70 ℃ [J].

Electrochim. Acta, 2006, 51: 3168-3172.

[40] Imran Jafri R, Sujatha N, Rajalakshmi N, et al. Au-MnO$_2$/MWNT and Au-ZnO/MWNT as oxygen reduction reaction electrocatalyst for polymer electrolyte membrane fuel cell[J]. Int. J. Hydrogen Energy, 2009, 34: 6371-6376.

[41] Pan C, Zhu Y. New type of BiPO$_4$ oxy-acid salt photocatalyst with high photocatalytic activity on degradation of dye[J]. Environ. Sci. Technol., 2010, 44: 5570-5574.

[42] Zhang A Y, Long L L, Liu C, et al. Electrochemical degradation of refractory pollutants using TiO$_2$ single crystals exposed by high-energy {001} facets[J]. Water Res., 2014, 66: 273-282.

[43] Sun Y, Liu Q, Gao S, et al. Pits confined in ultrathin cerium(IV) oxide for studying catalytic centers in carbon monoxide oxidation[J]. Nat. Commun. 2013, 4: 2899.

第 6 章

TiO_{2-x} 单晶电催化还原降解硝基苯

6.1 概述

电化学还原法是一种有效的污染物降解方法,尤其适用于含有吸电子基(如—NO_2 和—SO_3H)污染物的降解[1-7]。硝基苯(NB)是一种有毒的致癌污染物,对人体健康构成严重的威胁。由于—NO_2 具有很强的电子亲和能力,它在常用的氧化降解体系中表现出极强的抗氧化性和稳定性[3-4,7]。因此,NB 的阴极还原成为一种有效的降解方法[8-16]。在污染物的电化学还原过程中,阴极材料发挥着至关重要的作用[1]。目前为止,各种金属阴极材料已经被陆续研究报道[2-16]。贵金属(如 Pd、Pt 和 Ag 等)在低过电位下具有良好的催化还原活性,但由于成本高、丰度低,其实际应用受到严重限制[2-16]。而非贵金属(如 Fe、Cu、Ni 等)的活性相对较低,稳定性较差[2,10-16],因此也难以进行推广应用。与金属和双金属材料相比,过渡金属氧化物具有廉价、稳定和易于制备等优势[17-19]。然而,由于它们在氧化还原反应中通常表现出惰性[20],因而无法驱动具有一定过电位的阴极还原反应。过渡金属氧化物是一类重要的功能材料,如果能充分开发其高效电催化还原难降解污染物的功能性质,将会实现重要的技术突破。

利用还原型≡Ti(Ⅲ)或氧空位对 TiO_2 进行结构自修饰,能广泛提升材料的催化应用前景[26-29]。与理想晶体相比,经过精细修饰原子结构的缺陷型 TiO_2 含有更活泼的电子结构,具有更好的光学活性、催化活性和离解吸附性能。例如,TiO_2 的非化学计量还原可以显著提高电子供体密度和电导率,并通过将带隙降低至 2.0 eV 以下来降低电荷转移阻力[26]。在这种情况下,生成的≡Ti(Ⅲ)可以作为一个有效的电子供体,它的电子可以被激发到导带(或者临近的≡Ti(Ⅳ))[30]。同时,自掺杂的≡Ti(Ⅲ)可以显著增强 TiO_2 表面对水和质子的吸附和催化性能。例如,氢化 TiO_2 在室温、无光照条件下即可分解气态甲醛,研究发现其催化活性的提升与氧空位缺陷有关[31]。然而在本征状态下,纯 TiO_2 不具备催化还原活性。缺陷型的自掺杂 TiO_{2-x} 具有良好的抗氧化性能和催化稳定性[26-29]。例如,TiO_{2-x} 在空气中暴露超过一年之后,仍然保持原来的颜色。同时,在强氧化光催化作用下,其活性无明显降低,缺陷型≡Ti(Ⅲ)和氧空位稳定存在[26]。因此,研究 TiO_2 的晶体缺陷掺杂和设计,对其结构改善和催化应用具有重要的意义。

受 TiO_2 特异晶面和缺陷掺杂相关特性的启发,可以推测,含有缺陷态原子和电子结构的还原型 TiO_{2-x} 具备电催化还原性能。因此,在本章的工作中,首先制备并表征了高能{001}晶面暴露的缺陷型 TiO_{2-x} SCs,并将其作为阴极材料进行污染物硝基苯的电催化还原降解;然后,从转化效率、反应选择性和能耗等方面对 TiO_{2-x} 的 NB 还原性能进行系统评估,明晰催化机理;最后,通过各项表征手段,验证缺陷和晶面工程调控的 TiO_{2-x} 作为阴极催化剂的有效性与可行性。

6.2 TiO_{2-x} 单晶的阴极性能表征与电催化还原体系设计

6.2.1 TiO_{2-x} 单晶的结构表征

通过水热法制备合成了晶面调控的{001}-TiO_2 SCs[26]。在 H_2 气氛(5 vol% H_2 + 95 vol% Ar)中,将 TiO_2 SCs 分别在 300~800 ℃条件下(400 ℃为标准组)高温还原煅烧 5.0 h,制备出具有高能{001}晶面的缺陷型 TiO_{2-x} SCs。以低能{101}晶面(平均粒径为 25 nm,锐钛矿/金红石 = 80:20(质量比),比表面积约为 50 $m^2 \cdot g^{-1}$)暴露为主的 P25(图 6.1)作为对照[25]。以碳纸作为基底,采用滴涂法制备了 TiO_{2-x}/C 阴极,并在氮气气氛中煅烧烧结,以提高电极的机械强度和结晶度。同时以 K_2PdCl_6 溶液为前驱体,在恒定的阴极电流下,用电化学方法将 Pd 纳米粒子沉积在碳纸上,作为对照电极(Pd/C)[8]。

使用电化学工作站(CHI 760D, Chenhua Co, China)进行所有的电化学测试。电化学阻抗谱(EIS)以 5.0 mV 为偏压,扫描频率为 $10^5 \sim 10^{-2}$ Hz,反应体系为 10 $g \cdot L^{-1}$ NaCl 溶液,以玻碳电极作为工作电极(0.196 cm^2),以铂丝(纯度>99.0%)作为对电极,SCE 作为参比电极。在 0.1 $mol \cdot L^{-1}$ Na_2SO_4 水溶液、

0～1.0 V 电压范围和 1000 Hz 固定频率的条件下,采用阻抗法测量莫特-肖特基曲线。使用之前,GC 电极通过 0.05 μm 氧化铝抛光粉打磨抛光,然后在蒸馏水中超声清洗,每次实验前用纯氩气对电解液除氧 10 min。电化学析氢反应(Hydrogen Evolution Reaction,HER)和 CV 扫描分别在 0.5 mol·L^{-1} KOH 水溶液和 0.1 mol·L^{-1} Na$_2$SO$_4$ 水溶液中以 5.0 mmol·L^{-1}[Fe(CN)$_6$]$^{3-}$/[Fe(CN)$_6$]$^{4-}$ 为反应介质进行测定,扫速为 50 mV·s^{-1}。

6.2.2
硝基苯的电催化还原

研究工作以硝基苯作为目标污染物,反应温度为室温,反应体系为三电极单室电解池,工作阳极有效面积为 6 cm^2,TiO$_{2-x}$ SCs 负载量约为 0.05 mg·cm^{-2}(共 0.30 mg),使用面积相同的 Ti 片作为对电极、SCE 为参比电极。工作电极与对电极的间距设置为 1 cm。使用电化学工作站作为驱动电源,将 80 mL、含 0.1 mol·L^{-1} Na$_2$SO$_4$、不调整 pH 的 NB 溶液施加 -0.5～-1.3 V(SCE)偏压进行电解,转速为 500 r·min^{-1},每隔一段时间进行取样。在相同的条件下进行 TiO$_2$、P25、Pd/C 的阴极还原实验作为对照。所有实验均平行三次,算出平均值和标准差。循环实验时,用蒸馏水将催化剂反复彻底冲洗,随后在 60 ℃下干燥 2.0 h,重复使用。

6.2.3
辅助测试分析

NB 浓度通过高效液相色谱(HPLC-1100,Agilent Inc.,USA)来测定,通过 Hypersil-ODS 反相柱,VWD 检测波长为 254 nm,流动相为水和甲醇($V_\text{水}$/$V_\text{甲醇}$ = 40∶60),流速为 1.0 mL·min^{-1}。矿化效率由 TOC 分析仪(Vario TOC cube,Elementar Co.,Germany)来测定和计算。中间产物主要为亚硝基苯(NSB)、苯基羟胺(PHA)和苯胺(AN),通过液相质谱(LCMS-2010A,Shimadzu Co.,Japan)和气相质谱(GCMS Premier,Waters Inc.,USA)进行鉴定。

通过场发射扫描电子显微镜（FE-SEM，SIRION200，FEI Co.，the Netherlands）、高分辨透射电子显微镜和选择区电子衍射（HRTEM/SAED，JEM-2100，JEOL Co.，Japan）对其形貌和结构进行了表征，使用 Builder 4200 仪器（Tristar Ⅱ 3020M，Micromeritics Co.，USA）测量表面积，通过 X 射线衍射（XRD，X'Pert，PANalytical BV，the Netherlands）分析晶体结构，通过电子顺磁共振（ESR，JES-FA200，JEOL Co.，Japan）测定电子态，通过 UV/Vis 分光光度计（UV 2550，Shimadzu Co.，Japan）测量紫外-可见-红外漫反射光谱（DRS），通过 X 射线光电子能谱（XPS，PHI 5600，Perkin-Elmer Inc.，USA）测定化学成分。通过 FTIR（Magna-IR 750，Nicolet Instrument Co.，USA）测定红外光谱，扫描范围为 4000～400 cm^{-1}，溴化钾压片。通过水接触角分析仪（JC2000A，Powereach Co.，Shanghai，China）测定亲水性。

6.3 TiO_{2-x} 单晶电催化还原硝基苯的效能与机理分析

6.3.1 TiO_{2-x} 单晶的阴极性能表征

由图 6.1 可知，TiO_2 和 TiO_{2-x} 均为形貌规整均一的纳米片，平均尺寸为 80～100 nm，厚度为 20 nm 左右，比表面积为 14.0 m^2·g^{-1}。高分辨率透射电镜图像和选择性区域电子衍射图谱进一步证实，TiO_2 和 TiO_{2-x} 在 F$^-$ 作为选择性盖层试剂的作用下定向生长，均表现出高能量{001}晶面暴露[26-29]。透射电镜图像显示，在低煅烧还原温度（300～500 ℃）下得到的 TiO_{2-x} 晶体结构中出现了额外的空腔（图 6.1(b～f)），而未经过还原的本征催化剂没有这种空腔结构（图 6.1(a)）。在高煅烧还原温度下，由于氧的缺失，局域晶体结构发生原子坍塌，从而导致了腔体的随机分散。当还原温度进一步升高至600 ℃以上,催化剂逐渐转化为金红石相,TiO_2的晶体结构中出现明显的黑点(图6.1(g~i))[21]。

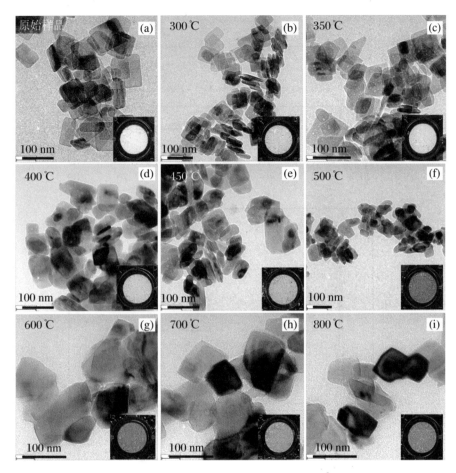

图6.1 不同煅烧温度下制备TiO_{2-x} SCs的TEM照片和颜色特征

X射线衍射图谱显示,TiO_2和TiO_{2-x}均为纯锐钛矿相,在600~800 ℃条件下制备的TiO_{2-x}中观察到金红石相的附加衍射峰[25]。这些结果与透射电镜图像一致,证实了在高温条件下发生的晶相转换过程(图6.2(a))。拉曼光谱表明TiO_{2-x}中存在氧空位,从改性后的几何结构和表面结构可以看出,特征峰发生了2.0~6.0 cm^{-1}的正偏移(图6.2(b))[26-29]。

XPS能谱进一步说明还原型的Ti^{3+}和氧空位大量出现在H_2煅烧还原的TiO_{2-x}催化剂中(图6.2(c、d))。此外,催化剂的可见光吸收也得到了不同程度的增强,这得益于缺陷活性位点的生成(Ti^{4-n}和氧空位)(图6.2(e))[30]。通过ESR进一步证实了这一观察结果(图6.2(f)),对应于$g=1.993$和$g=2.006$附近的强信号[26]。然而,由于暴露于大气中的表面氧化作用,在$g=2.02$时没有发现明显的Ti^{3+}表面信号。这些形态和结构的测量结果清楚地表明,还原

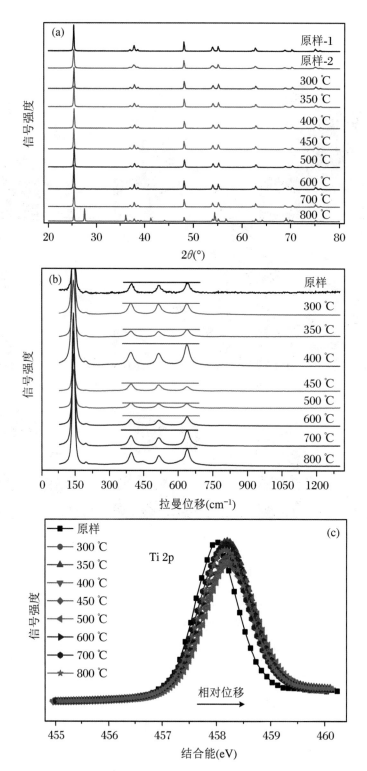

图 6.2 不同煅烧温度下制备的 TiO_{2-x} SCs 和本征 TiO_2 SCs 的结构特性

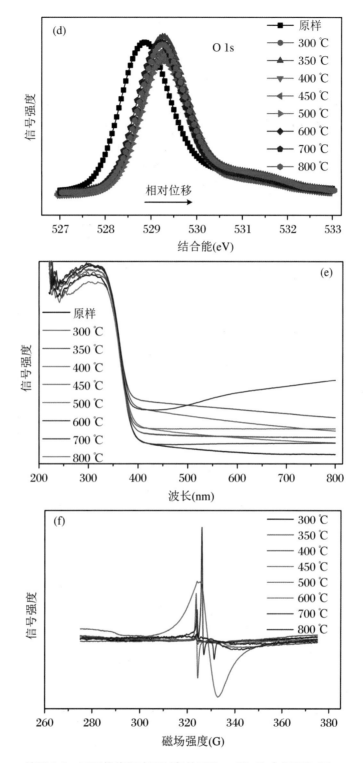

续图 6.2 不同煅烧温度下制备的 TiO_{2-x} SCs 和本征 TiO_2 SCs 的结构特性

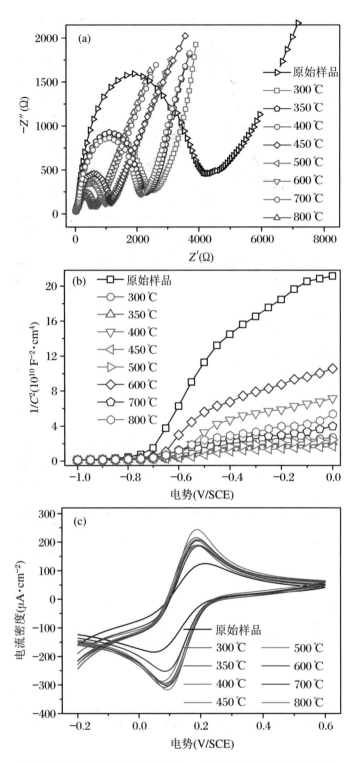

图 6.3　不同煅烧温度下制备 TiO_{2-x} SCs 的电化学测试和表面性能表征

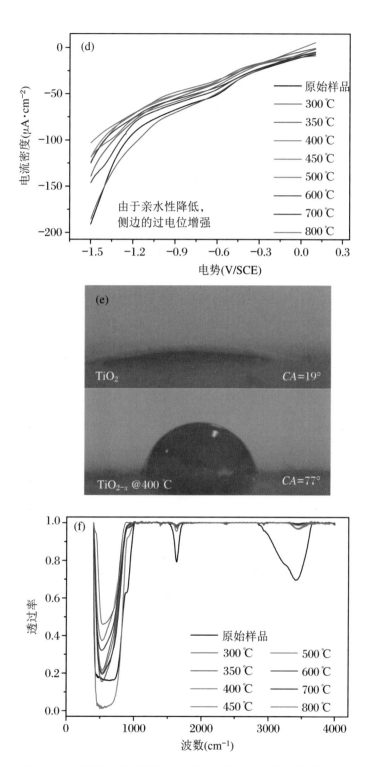

续图 6.3　不同煅烧温度下制备 TiO_{2-x} SCs 的电化学测试和表面性能表征

H_2 气氛热处理的晶体缺陷可以有效地修饰 TiO_2 SCs 的局部几何结构和电子结构。

引入氧空位/Ti^{3+} 后,TiO_{2-x} SCs 的局部几何结构和电子结构发生改变,这对其性能的改变和提升起着重要作用[26-29]。首先,由于电子给体密度大幅度增加(图 6.3(b)),电导率显著提高,电荷转移电阻显著降低,有利于氧化还原反应中的电子转移(图 6.3(a))。此外,利用[$Fe(CN)_6$]$^{3-}$/[$Fe(CN)_6$]$^{4-}$ 的氧化还原电对证实,当氧空位/Ti^{3+} 结构出现在表面和亚表面时,TiO_2 的电化学活性显著增强(图 6.3(c))。缺陷型 TiO_{2-x} 表现出明显更高的氧化还原峰和更窄的峰间距,说明它具有良好的电化学活性[25]。

在电化学污染物的还原过程中,阴极材料的 HER 过程起着非常重要的作用,越高的析氢过电位越有利于污染物的降解[8-16]。在相同条件下,TiO_{2-x} 的 HER 性能显著优于 TiO_2(图 6.3(d)和图 6.4),阴极电流越小,HER 作为副反应的活化能越高。TiO_{2-x} 的性能提升可能是由于亲水性相对 TiO_2 发生了变化,水接触角与 FTIR 测试进一步验证了这个猜测(图 6.3(e、f))。

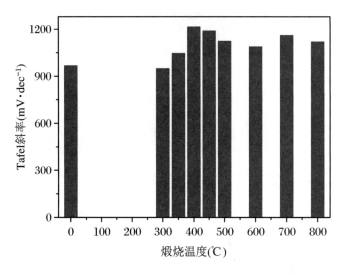

图 6.4 不同煅烧温度下制备 TiO_{2-x} SCs 的 HER 性能对比

之前的研究表明,TiO_2 经过 H_2 煅烧还原后,表面的 Ti—OH 键会被 Ti—H 部分取代[29]。显然,缺陷型 TiO_{2-x} 电荷转移电阻的降低、电化学活性和选择性的提高,使它成为一种有前途的还原型电催化材料。

6.3.2
电催化还原硝基苯的效能评估

通常情况下,NB 的电化学还原可以产生多种中间产物,其还原程度主要取决于 pH 和阴极材料性能[2-7]。在这些中间产物中,苯胺(AN)毒性最低、生物降解性能最好,因此将 NB 选择性地还原为 AN 具有重要的水处理意义。本工作中,含 $0.1\ mol·L^{-1}\ Na_2SO_4$ 的 $10\ mg·L^{-1}$ NB 水溶液(pH 未调控)在 GC、P25 和 TiO_2 电极上的还原电流很小,仅在约 $-0.60\ V/SCE$ 处出现还原峰(图 6.5(a))。相比之下,TiO_{2-x} 电极上的还原电流明显增加,并且在 $-1.15\ V/SCE$ 处有一个额外的还原峰(图 6.5(a))。此外,在缺陷型 TiO_{2-x} 电极上不同偏压下的电解证实了这一结果(图 6.5(b)),当施加的偏压低于 $-0.6\ V/SCE$ 时,NB 的去除率显著提高,而偏压为 $-0.5\ V/SCE$ 时,去除率小于 20%。这一结果可能主要归因于电辅助吸附,一种与任何电子转移无关的非共进过程[32-33]。此外,当阴极电位下降至比析氢电位更低的 $-1.3\ V/SCE$ 时,可以明显观察到析氢反应的发生,从而显著降低了电流效率。

在 CV 测试中,$-0.60\ V/SCE$ 处的还原峰对应 $4\ e^-$ 还原过程 NB→PHA,然后 PHA 进一步经过 $2\ e^-$ 还原过程生成苯胺,对应还原峰为 $-1.15\ V$。因此,可以推测出缺陷型 TiO_{2-x} SCs 对 NB 还原路径。此外,在 $-1.15\ V/SCE$ 处苯胺还原峰值比析氢加剧电位(约 $-1.30\ V/SCE$)更正,这表明,在缺陷型 TiO_{2-x} 上将 NB 电催化还原为 AN 的过程具有较高的电流效率和较低的能耗。

在 $0.1\ mol·L^{-1}\ Na_2SO_4$ 水溶液中(不调控 pH),以 $-1.2\ V/SCE$ 恒电位模式电解 NB 还原(图 6.5(c))。实时电流显示,降解反应在 100 s 内达到稳态,电流值随着 NB 的消耗而逐渐下降。

在缺陷型 TiO_{2-x} SCs 电极上,初始 1.0 h 内 NB 浓度迅速下降(去除率接近 90%),2.0 h 电解后去除率达到 100%。相比之下,在 P25、TiO_2 和碳纸电极上电解 1.0 h 后,NB 去除率仅为 60%、40% 和 20%。TiO_{2-x} 电极的 NB 去除率要高得多,一级动力学速率常数(图 6.5(d))和平均电流效率(图 6.5(e))也要大得多,表明氧空位/≡Ti(Ⅲ)发挥了关键作用。P25 对 NB 的电化学还原活性高于

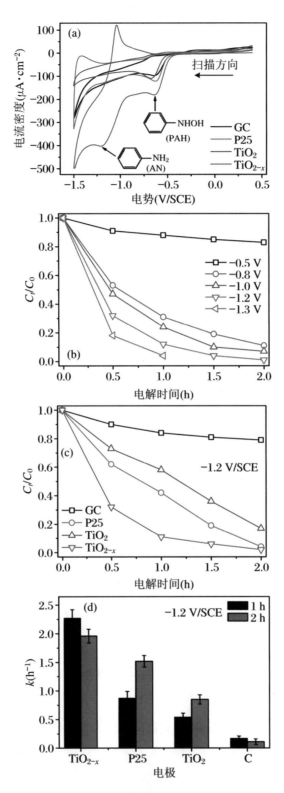

图 6.5 400 ℃ 煅烧制备的 TiO_{2-x} SCs 与对照材料对 NB 的电催化还原

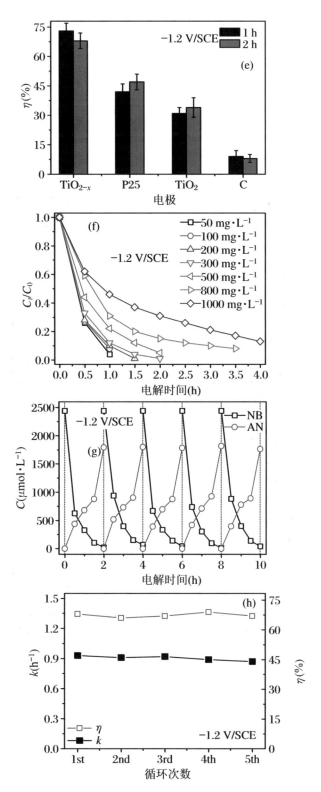

续图 6.5 400 ℃煅烧制备的 TiO_{2-x} SCs 与对照材料对 NB 的电催化还原

TiO$_2$(图 6.5(a、c~e)),这可能是由于混合相晶体结构存在缺陷,因而具有良好的给电子性能、较大的比表面积和较好的分散性所致。在相同条件下,TiO$_{2-x}$ SCs 还原 NB 的电化学活性甚至可以与贵金属 Pd/C 媲美(图 6.1 和图 6.6)[2],这种性能优势主要归功于其晶体表面和亚表面氧空位缺陷上所固定的较高浓度的原子氢·H。当 NB 浓度增加到 $1.0\,\mathrm{g \cdot L^{-1}}$ 时,在 4.0 h 时的去除率依旧保持在 80% 以上(图 6.5(f))。

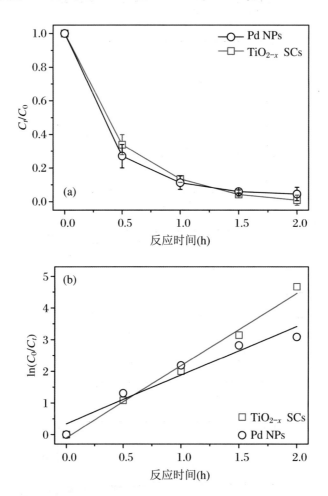

图 6.6 400 ℃ 煅烧制备的 TiO$_{2-x}$ SCs 与 Pd/C 电极的 NB 降解对比

电极稳定性是影响其应用的一个重要因素,一些典型金属由于其理化和催化稳定性差,使用寿命短,很难实际应用于水处理[1,11]。在 5 次循环降解实验中,NB 去除效率和产生量几乎保持不变,重复实验中各取样点的浓度值也基本持平,表明缺陷型 TiO$_{2-x}$ SCs 具有良好的稳定性和可靠性(图 6.5(g))。

一级动力学速率常数 k 为 0.9，电流效率为 68%，属于非常优异的阴极材料（图 6.5(h)）。

我们对比测定了缺陷型 TiO_{2-x} SCs 上 NB 还原主要中间产物的转化率（图 6.7）。在 TiO_{2-x}、TiO_2 和 P25 电极上，检测了 NSB、PHA 和 AN 三种典型中间产物，并通过其 CV 图（图 6.5(a)）进行了验证。此外，TiO_{2-x} 缺陷电极上还原中间产物积累量高、转化速率快，说明其具有较好的 NB 还原能力，主要产物为低毒的 AN（图 6.7(a)）。相比之下，在碳纸电极上只观察到 PHA（图 6.7(b)），表明其对 NB 还原的电化学活性较弱。

在 H_2 高温煅烧还原反应中，缺陷型 TiO_{2-x} SCs 中氧空位或还原型 $\equiv Ti(Ⅲ)$ 的含量随着煅烧温度的升高而逐渐增加（图 6.1）[26-29]。缺陷的引入固然大幅提升了 NB 的还原能力和阴极电流效率（图 6.8），但所有缺陷型 TiO_{2-x} SCs 整体上相对本征态 TiO_2 均表现出更好的催化还原性能，这意味着晶体缺陷形成、几何和电子调控对电化学活性提升具有重要意义。

6.3.3

缺陷中心电催化还原机理

在电化学加氢过程中，水的电解作用会使电极表面生成固定的原子氢[2]，原子氢的活性受阴极材料化学成分、元素价态、晶体结构和表面形态的影响[1,11]。缺陷型 TiO_{2-x} SCs 的阴极催化优势很可能归功于其所具有的单晶结构、高能 {001}晶面和 Ti^{3+}/氧空位（图 6.1）。电导率的显著提升、内部结构的连续有序和结晶度的增强，有利于加速催化过程中的电子转移、降低电极极化（图 6.3(c) 和图 6.3(d)）[23-24]。由于未配位高密度 Ti 原子（100% 5 配位的 Ti 原子，Ti_{5c}）的存在[23-24]，其晶格表面暴露的高能 {001} 晶面含有独特的原子、电子和高能结构，能在离解吸附和电荷迁移过程发挥介导作用（图 6.5 和表 6.1），从而使 TiO_{2-x} 具有优越的电化学活性[25]。

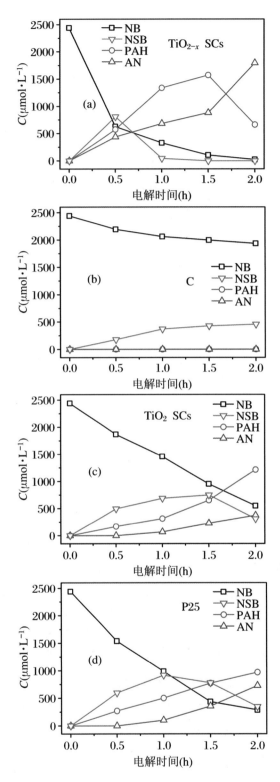

图 6.7 400 ℃ 煅烧制备的 TiO_{2-x} SCs 电化学还原 NB 的主要中间产物

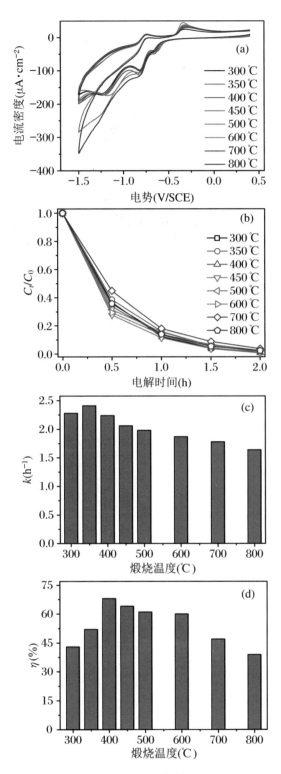

图 6.8　不同煅烧温度下制备的 TiO_{2-x} SCs 对 NB 的电化学催化还原

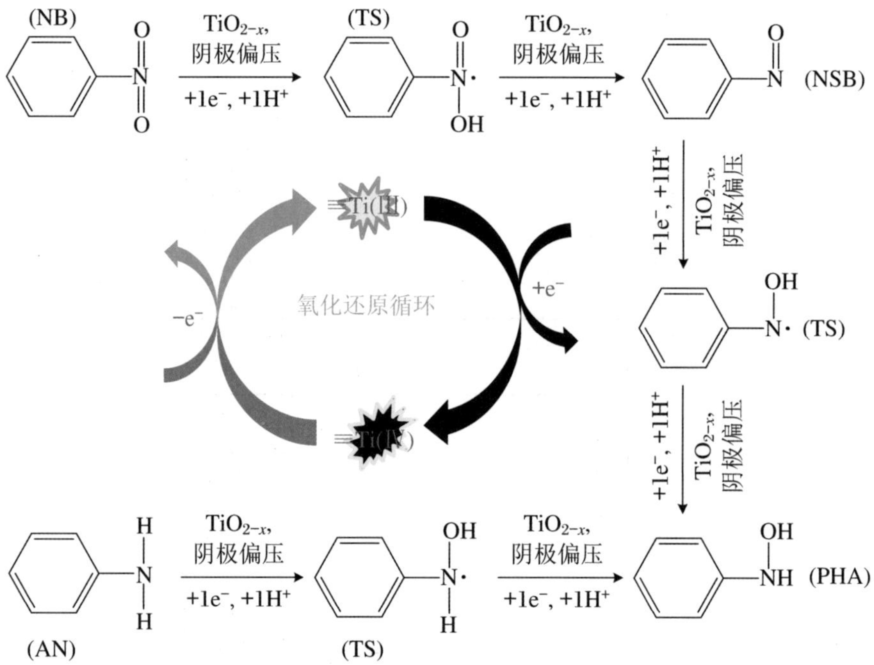

图 6.9 缺陷中心对 TiO_{2-x} 还原 NB 的催化反应路径的影响：氧空位区域钛的 +3～+4 混合价态循环，为还原电子从基底迁移到 TiO_{2-x}、再迁移到质子附近反应生成活性氢（·H/·H⁻）提供了一种有效的电子转移路径

表 6.1 TiO_{2-x} 的 Langmuir-Hinshelwood 一级速率常数（h-k）和电流效率（η,%）

反应条件		阴极电势（V/SCE）①					NB 初始浓度（mg·L⁻¹）②					
		−0.5	−0.8	−1.0	−1.2	−1.3	50	100	300	500	800	1000
TiO_{2-x}	k③	0.09	1.09	1.37	2.26	3.22	3.22	2.96	2.26	1.46	0.71	0.47
	η④	92.4	79.6	77.4	73.4	47.7	66.8	70.9	73.4	74.1	67.6	74.0
TiO_2	k③	0.07	0.34	0.49	0.85	1.75	0.71	0.78	0.85	1.20	1.43	1.59
	η④	93.4	64.7	60.3	34.1	56.0	30.8	30.4	34.1	41.6	40.9	38.3

注：① [NB] 为 300 mg·L⁻¹，NB 溶液体积为 80 mL，还原温度为 400 ℃，不调控 pH，搅拌速率为 500 r·min⁻¹，反应时间为 2.0 h。

② 阴极偏压为 −1.2 V/SCE，NB 反应溶液体积为 80 mL，还原温度为 400 ℃，不调控 pH，反应时间为 2.0～4.0 h。

③ $\ln(C_0/C_t) = kt$，C_0 $t=0$ 时刻的 NB 浓度，C_t $t=t$ 时刻的 NB 浓度。

④ $\eta = \dfrac{V(n_1 F C_{NSB} + n_2 F C_{PHA} + n_3 F C_{AN})}{Q} \times 100\%$

$= \dfrac{V(n_1 F C_{NSB} + n_2 F C_{PHA} + n_3 F C_{AN})}{\int_0^t I \mathrm{d}t} \times 100\%$。

其中，n_1、n_2 和 n_3 分别为 NB 还原生成三种中间产物的电子转移数：NSB（$n_1=2$），PHA（$n_2=4$），AN（$n_3=6$），V 为溶液体积（L），F 为法拉第常数，Q 为总电量（A·s），I 为电流（A），C_{NSB}、C_{PHA} 和 C_{AN} 分别为 NSB、PHA 和 AN 的浓度（mol·L⁻¹）。

更重要的是，氧空位的缺陷≡Ti(Ⅲ)可以作为 H_2O/H^+ 电解制氢的反应活性位点。在缺陷型 TiO_{2-x} SCs 中，体相或表面的氧原子缺失会形成一个或两个"自由"电子，和三个最近的钛原子可能发生进一步的键合作用[26-29]，并导致 TiO_2 在原子尺度上的几何电子结构优先暴露≡Ti(Ⅲ)的活性位点[20]。这种缺陷结构可能导致缺陷相关的性质，主要包括结构、电子、光学、离解吸附和还原等方面[26-29]。这些缺陷相关的特性使得 TiO_{2-x} 具有良好的潜力。一方面，缺陷≡Ti(Ⅲ)是活跃的、电子的和高度还原的，并且可以作为一种有效的氧化还原电对·H/·H$^-$，$Ti^{3+} + H^+ \rightarrow Ti^{4+} + \cdot H$ 或 $2Ti^{3+} + H^+ \rightarrow 2Ti^{4+} + \cdot H^{-}$ [26]；阴极电子很容易被束缚在缺陷型 TiO_{2-x} 的表面，可以减少 Ti^{3+} 转变为 Ti^{4+}。该过程受益于 Ti^{3+} 的形成[34]。另一方面，通过在氧空位处不同价态钛原子之间的电子转移，可以在锐钛矿晶体结构内可逆地氧化和还原缺陷≡Ti(Ⅲ)活性位点，而不受明显的本征晶体和电子结构约束[29]，H_2O/H^+ 的化学吸附和解离活化是一个关键调控步骤，但已有文献表明，这两种路径只适用 TiO_2 表面有缺陷的负电荷体系，而不适用于中性体系[26-29,35]。由于高催化活性的强耦合，H_2O/H^+ 在氧空位缺陷上的解离吸附可以导致更便捷的电子转移[23]。此外，TiO_{2-x} SCs 的高 HER 过电位表明，对于强反应性能的活性氢物种而言（即·H 和·H$^-$），其表面和亚表面的结合能都得到了明显改善（图 6.3(d)）。

$$\equiv Ti(Ⅲ) + H_2O \rightarrow \equiv Ti(Ⅲ) - H_2O^* \tag{6.1.1}$$

$$\equiv Ti(Ⅲ) + H^+ \rightarrow \equiv Ti(Ⅲ) - H^{+*} \tag{6.1.2}$$

$$\equiv Ti(Ⅲ) - H_2O^* \rightarrow \equiv Ti(Ⅳ) - \cdot H_2O^{-*} \rightarrow \equiv Ti(Ⅳ) - \cdot H^* + OH^- \tag{6.2.1}$$

$$\equiv Ti(Ⅲ) - H^{+*} \rightarrow \equiv Ti(Ⅳ) - \cdot H^* \tag{6.2.1}$$

$$\equiv C - OH + 阴极 \leftrightarrow \equiv C - OH^* - 阴极 \tag{6.3.1}$$

$$\equiv Ti(Ⅳ) - \cdot H^* + \equiv C - OH^* - 阴极 \rightarrow$$
$$\equiv Ti(Ⅳ) - H_2O^* / \equiv Ti(Ⅳ) - H^{+*} + \equiv C - H^* - 阴极 \tag{6.3.2}$$

$$\equiv C - H^* - 阴极 \leftrightarrow \equiv C - H + 阴极 \tag{6.3.3}$$

$$\equiv Ti(Ⅳ) - H_2O^* + e^- \rightarrow \equiv Ti(Ⅲ) - H_2O^* \tag{6.4.1}$$

$$\equiv Ti(Ⅳ) - H^{+*} + e^- \rightarrow \equiv Ti(Ⅲ) - H^{+*} \tag{6.4.2}$$

$$H_2O^* \rightarrow 1/2\, O_2 + 2H^+ + 2e^- \tag{6.5}$$

$$\equiv Ti(Ⅳ) - \cdot H^* + H_2O + e^- \rightarrow \equiv Ti(Ⅳ) - H_2O^* + H_2 \uparrow \tag{6.6.1}$$

$$\equiv Ti(Ⅳ) - \cdot H^* + \equiv Ti(Ⅳ) - \cdot H^* \rightarrow \equiv Ti(Ⅳ) - H_2O^* + H_2 \uparrow \tag{6.6.2}$$

在缺陷型 TiO_{2-x} SCs 表面，H_2O 和 H^+ 在电子转移（反应（6.1.1）和

(6.1.2))之前,首先聚集在具有高吸附能的≡Ti(Ⅲ)活性位点上,生成新的物种≡Ti(Ⅲ)—H$_2$O*/≡Ti(Ⅲ)—H^{+*}。它们只吸附在通过亲电吸附可以获得过量负电荷的位置[26-29,35]。通过吸附并被进一步活化的H$_2$O具有显著拉长的O—H键,可以捕获位于氧空位上的自由电子,同时生成≡Ti(Ⅳ)—·H$_2$O^{-*}的氢自由基(反应(6.2.1),前半部分反应)。这些自由基基团的形成,通过1/2电子路径(反应(6.2.1)~(6.3.3)),有效促进荷电分离和H$_2$O还原活化,生成高度还原的·H/H$^-$自由基。然后通过阳极氧化(反应(6.4.1)~反应(6.5))将≡Ti(Ⅲ)从≡Ti(Ⅳ)还原再生到另一个H$_2$O分子。但是,如果电化学加氢对于氢的解吸过程相比太慢(反应(6.6.1)和(6.6.2)),则只有析氢副反应发生[5]。在这个以缺陷位点为催化活性中心的反应(图6.10)中,活性≡Ti(Ⅲ)从氧化态≡Ti(Ⅳ)在碳阴极上被有效还原再生可能是电化学加氢的限速步骤。值得注意的是,由于电导率提高、电荷转移阻力降低和氧化还原晶体结构的多功能化,并且可以同时容纳TiO$_{2-x}$ SCs中还原态≡Ti(Ⅲ)和氧化态≡Ti(Ⅳ),因此在连续阴极反应中,≡Ti(Ⅲ)活性位点的原位再生可以被有效持续进行,而不受制于任何本征影响和电子结构约束。此外,污染物在基底上的电辅助吸附可以进一步促进以TiO$_{2-x}$为催化活性中心的电化学加氢[8,32-33]。

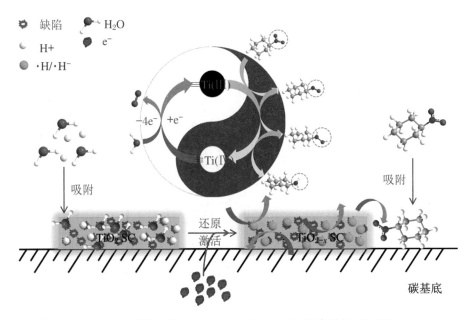

图6.10 以氧空位缺陷为催化活性中心的TiO$_{2-x}$对NB的电催化还原机理

为了进一步探索以活性原子H介导的NB间接电化学还原路径,选用400 ℃煅烧的TiO$_{2-x}$ SCs来进行叔丁醇抑制实验,以验证原子H的催化加成作用机理[36]。如图6.11所示,叔丁醇浓度从0增加到80.0 mmol·L^{-1}时NB去除率

C_t/C_0 及其 Langmuir-Hinshelwood 一级反应动力学常数 k_i 均大幅降低;当叔丁醇用量进一步增加时,催化活性基本保持不变。投加 80.0 mmol·L^{-1} 叔丁醇时,即叔丁醇/NB=20/1,k_i 下降到原始 NB 降解速率 k_0 的 31.5%(图 6.11)。这些结果表明,表面吸附的原子 H 是 NB 在 TiO$_{2-x}$/C 电极进行电化学还原的主要活性物种,间接电化学还原机理路径发挥了主导作用,只有少量 NB 通过电化学直接还原而降解[36]。

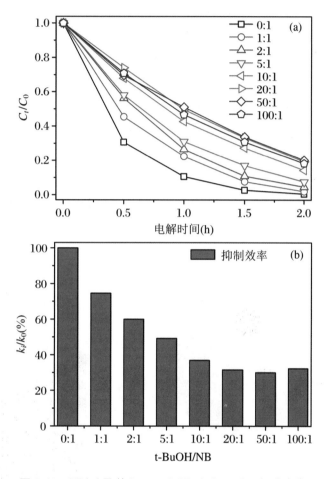

图 6.11　NB(a)及其 Langmuir-Hinshelwood 一级速率常数比(b)对不同 t-BuOH 浓度 0.0~30000 mg·L^{-1} 400 ℃ 制备的 TiO$_{2-x}$ SCs 的还原效率

本章小结

在本章工作中,借助晶面调控和缺陷调控策略,制备出表面和亚表面富含氧空位的缺陷型 TiO_2 单晶电极材料。所制备的 $\{001\}$-TiO_{2-x} 单晶作为一种优异的阴极电催化剂可高效、稳定地还原硝基苯(图 6.12);硝基苯在 $\{001\}$-TiO_{2-x} 单晶表面的电催化还原主要通过生成中间态高活性原子氢·H 完成;TiO_{2-x} 单晶优异的电化学活性主要缘于连续有序的晶体结构、活泼自由的电子结构和表面能极高的 $\{001\}$ 极性暴露晶面,而作为晶体缺陷位点的氧空位/$\equiv$Ti(Ⅲ)成为电子传递的催化活性中心;氧空位附近的表面低配位钛金属原子是主要催化活性位点,在+3、+4 价态之间实现快速、连续和稳定的热力学循环,进而完成原子氢的产生和硝基苯还原反应过程中的电子传递。高能$\{001\}$晶面暴露的缺陷型 TiO_{2-x} 单晶具有高活性、高稳定性、低成本和绿色无毒等优势,有望成为一种具有良好应用前景的新型高效电催化剂。

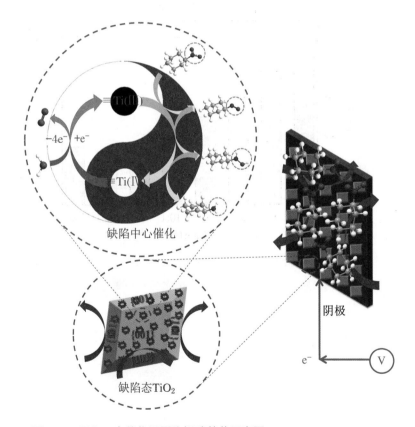

图 6.12　TiO_{2-x} 电催化还原降解硝基苯示意图

参考文献

[1] Rodrigo M A, Oturan N, Oturan M A. Electrochemically assisted remediation of pesticides in soils and water: a review[J]. Chem. Rev., 2014, 114: 8720-8745.

[2] Logue B A, Westall J C. Kinetics of reduction of nitrobenzene and carbon tetrachloride at an iron-oxide coated gold electrode[J]. Environ. Sci. Technol., 2003, 37: 2356-2362.

[3] Li Y P, Cao H B, Liu C M, et al. Electrochemical reduction of nitrobenzene at carbon nanotube electrode[J]. J. Hazard. Mater., 2007, 148: 158-163.

[4] Seshadri G, Kelber J A. A Study of the electrochemical reduction of nitrobenzene at molybdenum electrodes[J]. J. Electrochem. Soc., 1999, 146: 3762-3764.

[5] Noel M, Ravichandran C, Anantharaman P N. An electrochemical technique for the reduction of aromatic nitrocompounds in H_2SO_4 medium on thermally coated Ti/TiO_2 electrodes[J]. J. Appl. Electrochemi., 1995, 25: 690-698.

[6] Silvester D S, Wain A J, Aldous H L C, et al. Electrochemical reduction of nitrobenzene and 4-nitrophenol in the room temperature ionic liquid [C_4dmim][N(Tf)$_2$][J]. J. Electroanal. Chem., 2006, 596: 131-140.

[7] Chen Y, Li H Y, Liu W J, et al. Electrochemical degradation of nitrobenzene by anodic oxidation on the constructed TiO_2-NTs/SnO_2-Sb/PbO_2 electrode[J]. Chemosphere, 2014, 113: 48-55.

[8] Cheng I F, Fernando Q, Korte N. Electrochemical dechlorination of 4-chlorophenol to phenol[J]. Environ. Sci. Technol., 1997, 31: 1074-1078.

[9] Dabo P, Cyr A, Laplante F, et al. Electrocatalytic dehydrochlorination of pentachlorophenol to phenol or cyclohexanol[J]. Environ. Sci. Technol., 2000, 34: 1265-1268.

[10] Li T, Farrell J. Reductive dechlorination of trichloroethene and carbon tetrachloride using iron and palladized-iron cathodes[J]. Environ. Sci. Technol., 2000, 34: 173-179.

[11] Buddhika G, Naresh S, Peter S. Degradation of chlorinated phenols by zero valent iron and bimetals of iron: a review[J]. Environ. Eng. Res., 2011, 16: 187-203.

[12] Cwiertny D M, Bransfield S J, Lynn Roberts A. Influence of the oxidizing species on the reactivity of iron-based bimetallic reductants[J]. Environ. Sci. Technol., 2007, 41: 3734-3740.

[13] Agarwal S, Al-Abed S R, Dionysiou D D, et al. Reactivity of substituted chlorines and ensuing dechlorination pathways of select PCB congeners with Pd/Mg bimetallics[J]. Environ. Sci. Technol., 2008, 43: 915-921.

[14] Fang Y, Al-Abed S R. Correlation of 2-chlorobiphenyl dechlorination by Fe/Pd with iron corrosion at different pH[J]. Environ. Sci. Technol., 2008, 42: 6942-6948.

[15] Xie Y, Cwiertny D M. Chlorinated solvent transformation by palladized zerovalent iron: mechanistic insights from reductant loading studies and solvent kinetic isotope effects[J]. Environ. Sci. Technol., 2013, 47: 7940-7948.

[16] Han Y L, Yan W L. Bimetallic nickel-iron nanoparticles for groundwater decontamination: effect of groundwater constituents on surface deactivation [J]. Wat. Res., 2014, 66: 149-159.

[17] Panizza M, Cerisola G. Direct and mediated anodic oxidation of organic pollutants[J]. Chem. Rev., 2009, 109: 6541-6569.

[18] Yang Y, Li J X, Wang H, et al. An electrocatalytic membrane reactor with self-cleaning function for industrial wastewater treatment[J]. Angew. Chem., Int. Ed., 2011, 50: 2148-2150.

[19] Yang Y, Wang H, Li J, et al. Novel functionalized nano-TiO_2 loading electrocatalytic membrane for oily wastewater treatment[J]. Environ. Sci. Technol., 2012, 46: 6815-6821.

[20] Li Y H, Liu P F, Pan L F, et al. Local atomic structure modulations activate metal oxide as electrocatalyst for hydrogen evolution in acidic water [J]. Nat. Commun., 2015, 6: 8064.

[21] Chen X B, Mao S S. Titanium dioxide nanomaterials: synthesis, properties, modifications, and applications[J]. Chem. Rev., 2007, 107: 2891-2959.

[22] Chen A C, Holt-Hindle P. Platinum-based nanostructured materials: synthesis, properties, and applications[J]. Chem. Rev., 2010, 110:

3767-3804.

[23] Liu S G, Yu J G, Jaroniec M. Anatase TiO_2 with dominant high-energy {001} facets: synthesis, properties and applications[J]. Chem. Mater., 2011, 23: 4085-4093.

[24] Liu G, Yang H G, Pan J, et al. Titanium dioxide crystals with tailored facets[J]. Chem. Rev., 2014, 114: 9559-9612.

[25] Zhang A Y, Long L L, Liu C, et al. Electrochemical degradation of refractory pollutants using TiO_2 single crystals exposed by high-energy {001} facets[J]. Wat. Res., 2014, 66: 273-282.

[26] Pan X Y, Yang M Q, Fu X Z, et al. Defective TiO_2 with oxygen vacancies: synthesis, properties and photocatalytic applications[J]. Nanoscale, 2013, 5: 3601-3614.

[27] Liu L, Chen X B. Titanium dioxide nanomaterials: self-structural modifications[J]. Chem. Rev., 2014, 114: 9890-9918.

[28] Su J, Zou X X, Chen J S. Self-modification of titanium dioxide materials by Ti^{3+} and/or oxygen vacancies: new insights into defect chemistry of metal oxides[J]. RSC Adv., 2014, 4: 13979-13988.

[29] Chen X B, Liu L, Huang F Q. Black titanium dioxide (TiO_2) nanomaterials[J]. Chem. Soc. Rev., 2015, 44: 1861-1885.

[30] Pei D N, Gong L, Zhang A Y, et al. Defective titanium dioxide single crystals exposed by high-energy {001} facets for efficient oxygen reduction[J]. Nat. Commun., 2015, 6: 8696.

[31] Zeng L, Song W, Li M, et al. Catalytic oxidation of formaldehyde on surface of $H-TiO_2/H-C-TiO_2$ without light illumination at room temperature[J]. Appl. Catal., B, 2014, 147: 490-498.

[32] Li X N, Chen S, Quan X, et al. Enhanced adsorption of PFOA and PFOS on multiwalled carbon nanotubes under electrochemical assistance[J]. Environ. Sci. Technol., 2011, 45: 8498-8505.

[33] Wu M F, Jin Y N, Zhao G H, et al. Electrosorption-promoted photodegradation of opaque wastewater on a novel TiO_2/carbon aerogel electrode[J]. Environ. Sci. Technol., 2010, 44: 1780-1785.

[34] Ji Y F, Guo W, Chen H H, et al. Surface Ti^{3+}/Ti^{4+} redox shuttle enhancing photocatalytic H_2 production in ultrathin TiO_2 nanosheets/CdSe quantum dots[J]. J. Phys. Chem. C, 2015, 119: 27053-27059.

[35] Swaminathan J, Subbiah R, Singaram V. Defect-rich metallic titania (TiO$_{1.23}$): an efficient hydrogen evolution catalyst for electrochemical water splitting[J]. ACS Catal., 2016, 6: 2222-2229.

[36] Mao R, Li N, Lan H C, et al. Dechlorination of trichloroacetic acid using a noble metal-free graphene-Cu foam electrode via direct cathodic reduction and atomic H∗[J]. Environ. Sci. Technol., 2016, 50: 3829-3837.

第 7 章

分子印迹功能化 TiO_2 单晶电化学检测双酚 A

7.1 概述

环境内分泌干扰物对人类健康有着很大的威胁,因此我们在催化降解的同时,也希望能够对其进行原位检测,这需要快速、准确、灵敏的检测方法[1-2]。电化学传感技术因其准确性高、检测快速且方法简捷,而具有广阔的应用前景,典型的电化学传感器通常以化学/生物化学膜的电化学变换器作为导体材料[1]。然而,电化学传感器对某些污染物的响应很弱,这大大增加了直接检测的难度[1-2]。近年来,多种纳米结构碳材料(如C纳米管和石墨烯等)、贵金属(如Au、Pt、Pd等)及其复合物(如Pt/C纳米管、Pd/石墨烯等)、量子点(如CoTe等)被用作电化学检测材料[3-7];但是,碳纳米材料存在着金属杂质、异质、凝聚、不可逆吸附、弱响应和低灵敏度的缺陷[1],贵金属面临着高成本、低稳定性和储量稀缺的问题。因此,如何开发高效廉价的污染物电化学检测材料,成为亟待解决的问题。

与纳米碳和贵金属相比,过渡金属氧化物廉价、稳定且易于制备[8],然而它们往往呈现出氧化还原惰性,难以直接应用于电化学环境检测[8-9],本征态的TiO_2亦是如此[10-18]。经过改性调控的TiO_2单晶因其连续有序的晶体结构而具有良好的导电性[19-21],表面原子在高能{001}极化晶面上的排列和配位($0.90 J \cdot m^{-2}$)使其具有独特的化学稳定性、吸附性和催化反应性[20-23];但由于TiO_2的比表面积低、缺乏反应活性位点,在扩散控制的催化中表面吸附容量很低,故而难以进行高效的污染物检测。因此,要想充分开发TiO_2的电化学传感检测功能,需要对其界面吸附容量进行改善。

分子印迹(MI)技术通过在合适的固体基底上制造精巧的带着模板分子形状、大小和功能基团记忆的结合位点,提供了一种改善传感器的表面富集性能的有效而通用的方法[24-28]。通过利用该技术,模板的形状和功能能够转录到微孔材料上,其官能团构型可能记录在固体基底上[29-30]。最近,发展出一种直接的不涉及有机聚合物的无机框架分子印迹方法[31-33]。该技术通过基底和电极表面的高亲和性,形成了特定的分子印迹能力。此种结合由各种非共价相互作用,如H键、van der Waals力、π-π和静电效应驱动。与有机分子印迹技术相似,无机分子印迹技术用于选择性结合的主要化学识别组分也是锚定在无机基底上的表面官能团,通常是含氧基团,如—OH[31-33]。上述无机分子印迹技术具有很好的稳定性、高结合位点密度、有效的再生热力学、且无表面毒化。这些特点有利于分析物有效检测中的选择性聚集和快速分解[31-33]。

BPA 广泛存在于环境中，是危害较大的典型环境内分泌干扰物。BPA 的电活性苯基使它适用于电化学检测[33-35]。在本章的工作中，设计了一种新型的 TiO_2 基电化学传感器用于检测环境内分泌干扰物。合成了带有高能{001}暴露晶面和 BPA 分子印迹位点的 TiO_2 单晶（MI-TiO_2 SCs），并进行了其作为检测材料的表征和评估。TiO_2 的低导电率和催化活性通过形状和面设计得到提高，弱界面吸附容量通过直接无机分子印迹技术得到增强。同时也考察了 MI-TiO_2 SCs 检测水样和各种其他实际样品中 BPA 的适用性。

7.2
分子印迹功能化 MI-TiO_2 单晶的电化学传感性能表征

7.2.1
MI-TiO_2 单晶纳米片的制备方法

如图 7.1 所示，TiO_2 SCs 通过水热方法制备，将 TBOT 加入到 HF 溶液中（24 wt%）后在 180 ℃ 条件下反应 24 h。而 MI-TiO_2 SCs 的制备需要在 100 mL Teflon 高压釜中，将 25 mL TBOT、15 mL HF(24 wt%) 和 1.5～20.0 mg BPA 混合搅拌 0.5 h。冷却到室温后，先后经乙醇、水和 0.1 mol·L^{-1} NaOH 清洗，再经 60 ℃ 真空干燥 6.0 h 得到白色粉末，最后将其在 500 ℃ 空气中煅烧 2.0 h。

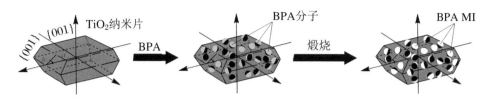

图7.1 BPA分子识别位点锚定在表面和亚表面的TiO₂ SCs的设计形成过程

7.2.2
MI-TiO₂ 单晶的材料表征

样品的形貌图由场发射扫描电子显微镜(FESEM)(JSM-6700F, JEOL Co., Japan)和透射电子显微镜(TEM)(JEM-2011, JEOL Co., Japan)拍摄。XRD 由 X 射线衍射仪(XPert, PANalytical BV, Netherlands)测量,激发源为 Cu Kα,波长 λ 为 0.154178 nm,单色器为石墨,散射角 2θ 为 10°～80°,扫速为 10°$\min^{-1}$。XPS 由光电子能谱仪(ESCALAB 250, Thermo VG, USA)测量,单色器激发源为 Ar,C 1s 的键能为 284.6 eV,作待测 Ti 和 O 的参照。FTIR 光谱在光谱仪(Vertex 70, Bruker Co., Germany)上记录,以 KBr 为基底。BET 和 BJH 分析通过自动比表面和孔径分析仪(Tristar Ⅱ 3020M, Micromeritics Co., USA)上完成(在液 N_2 环境下吸附-脱附平衡)。

7.2.3
电化学检测性能测试

所有电化学测试都于室温下在自制的 20 mL 三电极(Wuhan Gaussunion Co., China)池中使用电化学工作站(Shanghai CHI 760E, CH Ins., China)进行,工作电极为玻碳电极(GCE,直径为 3.0 mm),对电极为 Pt 丝,参比电极为 Ag/AgCl/饱和 KCl。体相溶液的 pH 由 pH 计(PHS-3C, Shanghai Leici Co., China)测量。部分测试在磷酸盐缓冲溶液(PBS)体系中进行。

改性前,GCE 先用 0.3 μm 和 0.05 μm Al_2O_3 粉末机械抛光,然后在去离子水中超声冲洗,以得到光滑表面。接着,在 0.5 mol·L^{-1} H_2SO_4 溶液中进行电化学活化处理，在 $-1.0\sim1.0$ V/SCE 之间以 100 mV·s^{-1} 的扫速进行 CV 测试，直到得到平稳的伏安曲线。随后，将 2.0 mg 催化剂在 2.0 mL 异丙醇和 4.0 μL Nafion 溶液中，超声 1.0 h，形成均匀分散液后，均匀滴加在抛光 GCE 上，并在空气中干燥。然后，在 5.0 mmol·L^{-1} K$_3$[Fe(CN)$_6$]∶K$_4$[Fe(CN)$_6$](1∶1) +1.0 mol·L^{-1} KCl 溶液中测试扫 CV 和电化学阻抗谱（EIS），在 0.1 mmol·L^{-1} K$_3$[Fe(CN)$_6$]+0.1 mol·L^{-1} KCl 溶液和 20 μm BPA+0.1 mol·L^{-1} PBS (pH=7.0) 溶液中进行计时库仑（CC, Q-t）分析。在 10 nmol·L$^{-1}\sim$20 μm BPA+0.1 mol·L^{-1} PBS(pH=3.0$\sim$13.0) 溶液中进行差分脉冲伏安（DPV）测试。通过十次空白溶液的标准偏差估算检测限（LOD）。

电极的电化学活性面积（A_{eff}）通过在模式溶液 0.1 mmol·L^{-1} K$_3$[Fe(CN)]$_6$ 和目标溶液 20 μm BPA 中的计时库仑法（CC, Q-t）测量，并根据 Anson 方程计算得到

$$Q(t) = \frac{2nFA_{\text{eff}}CD^{1/2}t^{1/2}}{\pi^{1/2}} + Q_{\text{dl}} + Q_{\text{ads}} \tag{7.1}$$

$$Q_{\text{ads}} = nFA_{\text{eff}}\Gamma_s \tag{7.2}$$

其中，Q 为总电量，n 为反应转移电子数，F 为法拉第常数（96485 C·mol^{-1}），C 和 D 分别为底物的浓度和扩散系数，Q_{dl} 和 Q_{ads} 分别为双电层电量和 Faradic 电量，Γ_s 为吸附容量。在模式溶液中，发生可逆反应，$n=1$，$C=0.1$ mmol·L^{-1}，$D=7.6\times10^{-6}$ cm^2·s^{-1}。

对于吸附控制并完全可逆的电极过程，氧化峰电势（E_p）由如下方程确定：

$$E_p = E_0 + \frac{2.303RT}{\alpha nF}\lg\left(\frac{RTk_0}{\alpha nF}\right) + \frac{2.303RT}{\alpha nF}\lg v \tag{7.3}$$

其中，E_0 为标准氧化还原电势，α 为电荷转移系数，R 为摩尔气体常数（8.315 J·mol^{-1}·K^{-1}），T 为热力学温度（298.15 K），k_0 为标准反应速率常数，v 为扫速。

将改性 GCE 在 5 μmol·L^{-1}、10 μmol·L^{-1} 和 20 μmol·L^{-1} BPA 溶液中，进行计时电流（CA, I-t）测量。不同浓度的 BPA 溶液中的 I-$t^{1/2}$ 曲线呈现为不同斜率的直线。从 I-t 得到的斜率，根据 Cottrell 方程，能够计算出特定的扩散系数

$$I(t) = \frac{nFACD^{1/2}}{\pi^{1/2}t^{1/2}} \tag{7.4}$$

其中，I 为法拉第电流，n 为反应转移电子数（$n=2$），A 为工作电极几何面积（0.196 cm^2），C 和 D 分别为 BPA 的浓度（mol·cm^{-3}）和扩散系数（cm^2·s^{-1}）。

初始表面浓度（Γ）表明 BPA 在阳极的吸附容量，由在不同电势扫速 v 下

BPA 在 MI-TiO$_2$/GCE 阳极的 LSV 曲线得到。当电势扫速增加，BPA 的氧化峰移向更正电势，观察到 E_p 随 $\lg v$ 的线性变化，证明了 BPA 的电化学氧化是不可逆的。

在吸附物质的不可逆电化学反应的情况下，能被表示为如下方程：

$$I_p = \frac{\alpha n n_a F^2 A v \Gamma}{2.718 RT} \tag{7.5}$$

其中，I_p 为峰电流，α 为电荷转移系数（通常被认为是 0.5），n 为电化学过程的总电子交换数（$n=2$），n_a 为在限制电子转移步骤的电子交换数（$n_a=2$），A 为电极面积（0.196 cm^2），R 为摩尔气体常数，T 为热力学温度（298.15 K），v 为扫速，Γ 为初始表面浓度。

因此，Γ 能够根据如下方程由 I_p 随 v 线性变化的斜率计算得到：

$$\Gamma = \frac{2.718 RT [Slope]}{\alpha n n_a F^2 A} \tag{7.6}$$

7.2.4
样品采集与预处理方法

所有的实际水样在分析前进行 0.45 μm 膜过滤。自来水样从本实验室采集，未经二次净化；湖水在合肥市的巢湖（中国的几大淡水湖之一）的不同位置采集；河水在合肥市的南淝河的不同位置采集；污水和污泥样品从二次出水（生物处理后、氯化消毒前）采集。湖水、河水和污水样品经过离心预处理，上清液用于 BPA 分析。污泥样品从二次沉淀池采集，并和 50 ℃ 30 mL 0.1 mol·L^{-1} PBS (pH=7.0) 混合，置于 30 ℃ 摇床振荡 1～5 天，随后离心去除沉淀，取上清液用于 BPA 分析。所有的塑料产品经过洗涤、干燥和切割，并在装有 50 ℃ 30 mL 0.1 mol·L^{-1} PBS(pH=7.0) 的密封锥形瓶中 30 ℃ 振荡 1～5 天浸出[36]。

7.3

MI-TiO$_2$ 单晶电化学检测双酚 A 的效能与机理分析

7.3.1

MI-TiO$_2$ 单晶的形貌与结构

原始和分子印迹 TiO$_2$ SCs 形状都是纳米片,长度为 50～80 nm,厚度为 5～10 nm。二者暴露{001}晶面的比例都超过 80%(图 7.2)。掺杂少于 5.0 mg BPA 作为模板,TiO$_2$ SCs 的形貌没有明显变化。三种样品的所有主峰都符合标

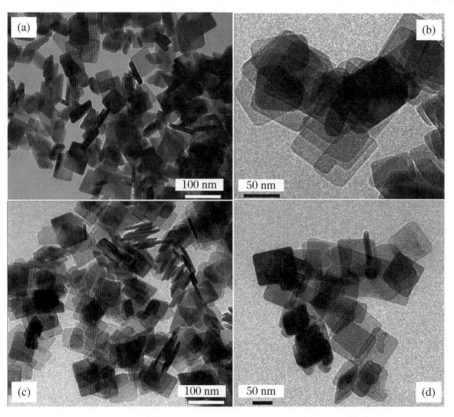

图 7.2　TiO$_2$ SCs(a、b)和 5.0 mg BPA 用量的 MI-TiO$_2$ SCs(c、d)的 TEM

准锐钛矿型(No. 78-2486)(图 7.3(a))。煅烧后 TiO$_2$ SCs 衍射峰变强和变锐[31]。TiO$_2$ SCs 的形貌与结构性质在模板添加和去除后保持不变(图 7.2)。分子印迹位点在 TiO$_2$ SCs 表面和亚表面形成,而非晶格间隙[31]。

FTIR 光谱为 BPA 的插入和去除后在 TiO$_2$ 表面和亚表面构建特定分子印迹位点提供了直接证据(图 7.3(b))。主峰在 1598 cm^{-1} 和 1510 cm^{-1}。芳香环上的 C—O—H 伸缩模式在 1237 cm^{-1}、1218 cm^{-1}、1117 cm^{-1}。C—H 的弯曲模式在 827 cm^{-1},—CH$_3$ 在 1421 cm^{-1}。这些峰清楚证明了 BPA 在 TiO$_2$ 基底中的存在及其煅烧后的完全消失[31]。O—H 在 3340 cm^{-1} 附近的伸缩模式和配位 Ti—OH 在 1612 cm^{-1} 的弯曲模式证明了丰富的活性基团的存在,它们通过 H 键相互作用与 BPA 特异性结合[31]。在 TiO$_2$ 表面检测到丰富的—OH,它们具有多个与 BPA 相互作用的位点,构建了印迹特异性结合空腔。同时,H 键和静电相互作用也记录了 BPA 的形状和构型[24]。而且,相比原始 TiO$_2$,MI-TiO$_2$ 的表面更粗糙、表面积更大,电化学识别能力更强(表 7.1)。

表 7.1 TiO$_2$ 电极材料的主要物理结构参数

材料	A_{BET}(m^2·g)	V_{Pore}(cm^3·g)	D_{Pore}(nm)
MI-TiO$_2$ SCs	55.0	0.219	15.97
TiO$_2$ SCs	36.9	0.071	7.71
P25	47.0	0.196	16.38

注:三次平行测量的平均值($n=3$),RSD 小于 10.0%。

7.3.2
MI-TiO$_2$ 单晶的电化学氧化活性

MI-TiO$_2$ 的氧化还原峰电势差(ΔE_p)得到大幅降低,为 120 mV,表明其阻抗最小,能够加速电子转移。再者,在三种 TiO$_2$ 材料中,MI-TiO$_2$ 的电化学活性面积最大。

在三种 TiO$_2$ 材料中,MI-TiO$_2$ 的电流最高、过电势最低(图 7.4(a))。相比 P25,原始 TiO$_2$ 的电流更小、过电势更低,表明面设计的 TiO$_2$ SCs 对于 BPA 氧化的活化能更低。并且,原始 TiO$_2$ 的表面吸附能较弱,限制了它整体反应速率。峰电流随着扫速增加而线性增加(图 7.4(b,c)),表明吸附在 BPA 电化学氧化中发挥作用。峰电势与电势扫速的对数线性相关,斜率为 $2.303RT/(\alpha nF)$(图

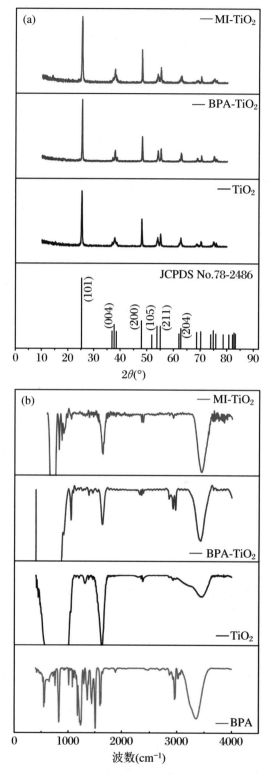

图7.3 TiO$_2$ SCs、5.0 mg BPA 用量的 BPA-TiO$_2$ SCs 和 MI-TiO$_2$ SCs 的 XRD(a) 和 FTIR(b)

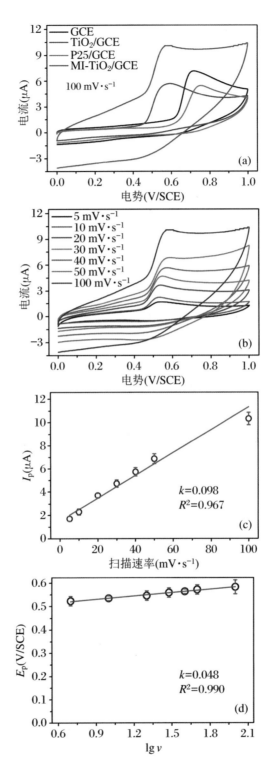

图 7.4　GCE、P25/GCE、TiO$_2$/GCE 和 MI-TiO$_2$/GCE 的 CV 曲线(a、b)及其 I_p-V(c)和 E_p-lg v(d)关系图

7.4（d）[36-37]。因而，MI-TiO$_2$/GCE 上 BPA 氧化反应的电子转移数约为 2[37-38]。

MI-TiO$_2$/GCE 上 BPA 的扩散系数随其表面浓度增加而显著增加，表明 MI-TiO$_2$ 与 BPA 之间的强结合作用源于其特异性分子识别能力。MI-TiO$_2$ 上优越的 BPA 氧化性能可归因为其很大的电化学活性表面和突出的表面聚集能力。此外，沉积的 MI-TiO$_2$ 用量并没有在很大程度上影响 BPA 的 DPV 检测。当 pH 从 3.0 增加到 6.0 时，氧化电流增加，在 pH=7.0 时维持在较高水平。随后，当 pH 从 8.0 增加到 13.0 时，氧化电流降低（图 7.5）。强酸溶液中，因其强质子化导致的强静电吸引，BPA 在 TiO$_2$ 上的吸附衰减；反之，在强碱性条件下，吸附增强[37-38]。最高电流对应的 pH 低于 BPA 酸电离常数（pK_a=9.73）[36]，证明了质子化的 BPA 在带负电的 MI-TiO$_2$ 上有更好的静电吸附（pH≈6.5）。

BPA 氧化的电势随着 pH 从 3.0 增加到 8.0 而降低，随后稳定在 pH=13.0。在酸性溶液中，观察到随着 pH 增加，E_p 朝着负电势方向线性移动，证明了质子的参与。其理论斜率（0.0561 V·pH^{-1}）与电子相近，表明酸性反应中的电子转移伴随着相同数量的质子[37-38]。没有观察到还原峰，表明 BPA 氧化的不可逆性[36]。由于该反应可以被描述为两电子过程（图 7.4（d）），MI-TiO$_2$ 上的 BPA 氧化也可以被认为是两电子-两质子路径[37-38]。此外，电辅助预吸附并没有对 MI-TiO$_2$/GCE 上的 BPA 检测起到促进作用，进一步证明了 MI-TiO$_2$ SCs 对 BPA 的特异性识别能力和强结合能力[31-32]。本例中，异相 BPA 氧化没有动力学限制。

7.3.3

双酚 A 的电化学检测性能评估

图 7.6(a)显示了在优化条件下 MI-TiO$_2$ 对不同用量的 BPA 的 DPV 响应。图 7.6(b)阐明了氧化峰电流与 BPA 浓度的校正曲线。从 10.0 nmol·L^{-1} 到 20.0 μmol·L^{-1}，氧化峰电流值与 BPA 用量成比例关系，相关系数为 0.9987。计算得到 LOD 为 3.0 nmol·L^{-1}（S/N=3），LOQ 为 10.0 nmol·L^{-1}（S/N=10）。运用 MI-TiO$_2$ 策略增强了分析信号、降低了 LOD、并扩展了 BPA 检测的线性范围（表 7.2）[39-74]。虽然一些其他材料达到相同甚至更好的检测性能[75-77]，但是 MI-TiO$_2$ 由于高活性、低成本、无毒性、高稳定性和丰富的地球储量，在电化学检测中具备更明显的优势。

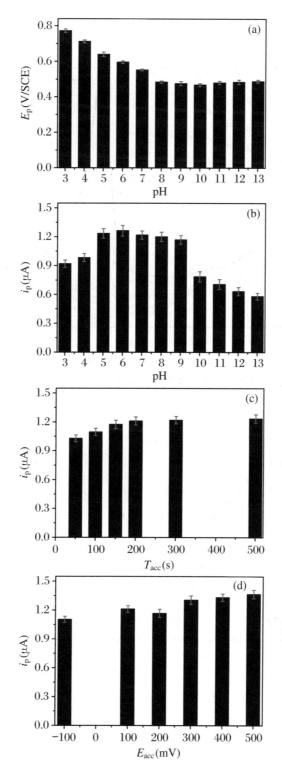

图 7.5 Ml-TiO$_2$/GCE 在不同 pH 条件下的 E_p(a)和 I_p(b),聚集电压为 100 mV、不同聚集时间(c)和聚集时间为 100 s、不同聚集电压(d)的条件下的 I_p

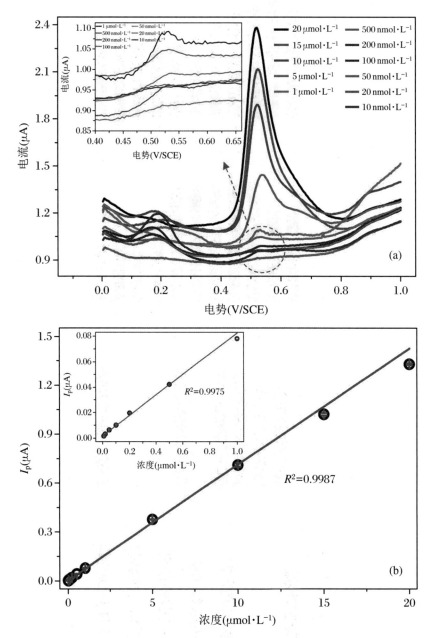

图7.6 MI-TiO$_2$/GCE 对没有额外的阳极聚集的不同浓度的 BPA 的 DPV 响应(a)、校正曲线(b)及其 I_p-v(c)和 E_p-lg v(d)关系图

表 7.2 本工作与参考文献中 BPA 检测方法的对比

电极	线性范围 (μmol·L^{-1})	LOD (S/N = 3, nmol·L^{-1})	方法	文献
/	2.19~219.29	170	LC-UV	[39]
/	0.0004~0.4000	0.08	ESI-LC-MS	[40]
/	0.0004~0.2193	0.04	GC-MS	[41]
/	79~16600	70000	FL	[42]
/	0.8~12.0	310	CL	[43]
CNTs/GCE	0.3~100.0	98	LC-Amperometry	[44]
N-GS	0.01~1.30	5	Amperometry	[45]
tyrosinase-G/CPE	0.1~15.0	100	Amperometry	[46]
tyrosinase-SWCNTs/CPE	0.1~12.0	20	Amperometry	[46]
tyrosinase-MWCNTs/CPE	1.0~16.0	1000	Amperometry	[46]
thionine/CPE	0.15~45.00	150	Amperometry	[47]
PAMAM-CoTe QDs/GCE	0.013~9.890	1	Amperometry	[48]
PAMAM-Fe$_3$O$_4$ NPs/GCE	0.01~3.07	5	Amperometry	[49]
PEDOT/GCE	90.0~410.0	22000	Amperometry	[50]
β-CD-SWCNTs/GCE	0.018~18.500	1	Amperometry	[51]
MAM-MWCNT/GCE	0.01~40.80	5	Amperometry	[52]
MWCNTs/GCE	0.05~20.00	20	Amperometry	[53]
NiTTPS-MWCNTs/GCE	0.05~50.00	15000	Amperometry	[54]
COOH-MWCNTs/GCE	0.001~10.000	5	LSV	[55]
PAM-MWCNTs/GCE	0.005~20.000	1.7	LSV	[56]
PGA-NH$_2$-MWCNTs/GCE	0.1~10.0	20	DPV	[57]
ITO	5.0~120.0	290	DPV	[58]
CPE	0.025~1.000	7.5	DPV	[59]
CoPc/CPE	0.0875~12.5000	10	DPV	[60]
MCM-41/CPE	0.22~8.80	38	DPV	[61]
CNF/CPE	0.8~50.0	100	DPV	[62]
N-CNF/CPE	0.1~60.0	50	DPV	[62]
G/GCE	0.05~1.00	46.9	DPV	[63]
Pt-G-CNTs/GCE	0.06~10.00, 10.00~80.00	42	DPV	[64]
Au-Pd NPs-G/GCE	0.01~5.00	4	DPV	[65]
chitosan-Fe$_3$O$_4$ NPs/GCE	0.05~30.00	8	DPV	[66]

续表

电极	线性范围 ($\mu mol \cdot L^{-1}$)	LOD (S/N = 3, $nmol \cdot L^{-1}$)	方法	文献
chitosan-Fe-rGO/GCE	0.06~11.00	17	DPV	[67]
chitosan-G/CILE	0.1~800.0	26.4	DPV	[67]
MWCNTs-Au NPs/GCE	0.02~20.00	7.5	DPV	[68]
Mg-Al-CO_3 LDH/GCE	0.01~1.05	5	DPV	[69]
tyrosinase-MWCNTs-CoPc-SF/GCE	0.05~3.00	30	DPV	[70]
pretreated BDD	0.44~5.20	210	DPV	[71]
pretreated pencil C	0.05~5.00, 5.00~10.00	3.1	DPV	[72]
rGO-CNTs-Au NPs/SPE	0.00145~0.02, 0.02~1.49	0.8	DPV	[73]
ZnO-CNTs/CILE	0.002~700.000	9	SWV	[74]
MI-TiO_2 SCs	0.01~20.00	3.0	DPV	本文

具有相似分子结构和化学性质的化合物，即使在100倍用量时，对于 MI-TiO_2 检测 BPA 的干扰也都可以忽略(图7.7(a))。该出色的 BPA 区分性可能归因于目标 BPA 与 TiO_2 表面和亚表面的分子识别位点之间的特异且强大的结合相互作用[31]。此外，其也表现出很高的检测稳定性。在4℃冰箱中储存10天、20天和30天后，检测电流分别保留了98.7%、98.0%和97.8%(图7.7(b))。加入0.5 $\mu mol \cdot L^{-1}$ 和1.0 $\mu mol \cdot L^{-1}$ BPA 后，检测电流分别增加至102.6%和104.3%，表明了良好的回收性(图7.7(c))。并且，其相对标准偏差很低(0.9%)，表现了优异的重现性(图7.7(d))。

对于表面参与的阳极反应，从基质到电极的直接电子转移只在预吸附在电极表面发生，并且质量传递是电化学反应的主要限速步。因而，BPA 的表面富集性能在其抗干扰能力中起着支配作用(图7.7)。线性扫描伏安法(LSV)测试显示出 MI-TiO_2 表面的初始 BPA 浓度，即使在严重干扰条件下，也能维持稳定。这很好地解释了为什么 MI-TiO_2/GCE 具有很高的抗干扰能力(表7.3)。

在 BPA 的电化学检测中引入了更多高浓度的分子尺寸或大或小的干扰物质。在给定条件下，MI-TiO_2 基传感器维持了相对稳定的 BPA 检测信号(降低少于10%)(表7.3)。进一步开展了四种真实结构类似物的对 BPA 分析的补充干扰测试。证明选择的这四种结构类似物对 MI-TiO_2 基传感器检测 BPA 也没有明显的干扰。即使在多重干扰条件下，分析信号仍大于95%，证明了BPA在

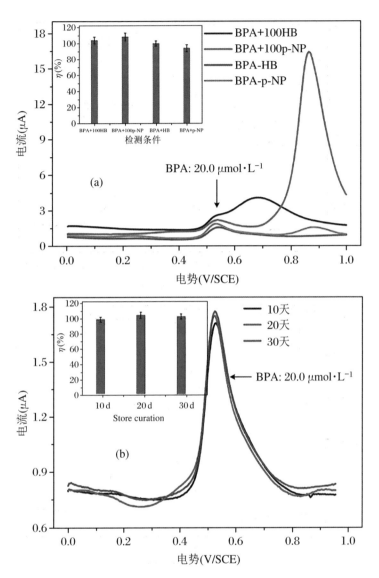

图 7.7　MI-TiO$_2$/GCE 电化学检测 BPA 的效能:选择性(a)、稳定性(b)、表观回收因子(c)和重现性(d)

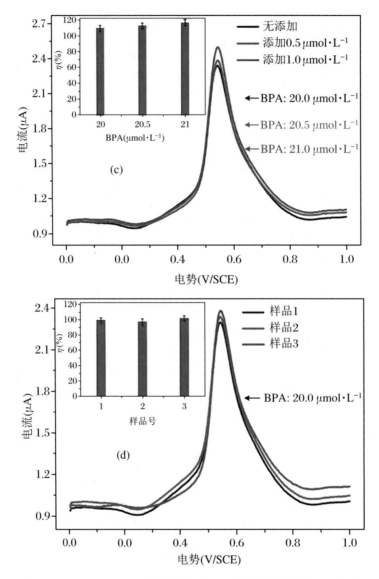

续图 7.7　MI-TiO$_2$/GCE 电化学检测 BPA 的效能：选择性(a)、稳定性(b)、表观回收因子(c)和重现性(d)

表 7.3　MI-TiO$_2$/GCE 检测 BPA 的初始表面浓度、峰电流和信号保留率[①]

分析条件	Γ ($\times 10^{-10}$ mol·L^{-1}·cm^{-2})	BPA 检测	
		i (μA)	η[②] (%)
BPA	9.83±0.37	1.218±0.041	100.00±3.37
BPA + HB	8.67±0.27	1.233±0.044	101.23±3.61
BPA + p-NP	9.02±0.31	1.151±0.050	94.50±4.11
BPA + p-NA	10.48±0.34	1.222±0.043	100.33±3.53
BPA + HA	9.39±0.23	1.213±0.052	99.59±4.27
BPA + CH$_3$OH	11.14±0.37	1.228±0.037	100.82±3.04
BPA + C$_2$H$_5$OH	9.28±0.39	1.203±0.035	98.77±2.87
BPA + HB + p-NP	8.40±0.42	1.135±0.045	93.19±3.69
BPA + HB + p-NA	7.19±0.31	1.205±0.044	98.93±3.61
BPA + p-NP + p-NA	10.52±0.35	1.231±0.039	101.07±3.20
BPA + HB + HA	9.03±0.49	1.104±0.038	90.64±3.12
BPA + p-NP + HA	8.15±0.36	1.273±0.041	104.52±3.37
BPA + p-NA + HA	9.27±0.35	1.199±0.029	98.44±2.38
BPA + HB + p-NP + p-NA	10.59±0.29	1.101±0.034	90.39±2.79
BPA + p-NP + p-NA + HA	7.45±0.28	1.221±0.042	100.25±3.45
BPA + HB + p-NP + HA	8.02±0.34	1.195±0.045	98.11±3.69
BPA + HB + p-NA + HA	8.48±0.35	1.222±0.044	100.33±3.61
BPA + HB + p-NP + p-NA + HA	7.07±0.26	0.925±0.038	75.94±3.12
BPA + 100HB	6.53±0.29	1.273±0.052	104.52±4.27
BPA + 100p-NP	8.69±0.29	1.325±0.063	108.78±5.17
BPA + 100HB + 100p-NP	11.14±0.38	1.311±0.051	107.64±4.19
BPA + 100HB + 100p-NA	10.39±0.34	1.206±0.036	99.01±2.96
BPA + 100HB + 100HA	9.01±0.54	1.141±0.042	93.68±3.45
BPA + 100p-NP + 100p-NA	7.46±0.51	1.177±0.047	96.63±3.86
BPA + 100p-NP + 100HA	8.20±0.39	1.099±0.039	90.23±3.20
BPA + 100HB + 100p-NP + 100p-NA	10.38±0.42	1.239±0.037	101.72±3.04
BPA + 100p-NP + 100p-NA + 100HA	7.04±0.38	1.132±0.035	92.94±2.87
BPA + 100HB + 100p-NP + 100HA	8.29±0.52	1.263±0.057	103.69±4.68
BPA + 100HB + 100p-NA + 100HA	7.46±0.51	1.164±0.052	95.57±4.27
BPA + 100HB + 100p-NP + 100p-NA + 100HA	9.01±0.34	1.219±0.036	100.08±2.96
BPA + HB + p-NP + p-NA + HA + CH$_3$OH + C$_2$H$_5$OH	6.38±0.44	0.831±0.042	68.23±3.45

注：[①] 取五次平行测量的平均值，RSD 小于 5.0%。
　　[②] $\eta = i/i_0 \times 100\%$，$i$ 和 i_0 分别为有干扰和无干扰时的检测电流。

TiO_2 上的特异性识别和选择性富集。考虑到水体中普遍存在的腐殖酸（HA）和 Cl^-、HCO_3^-、Fe、Mn，我们也选择用它们来模拟真实水体进行 BPA 分析。同样，它们对 MI-TiO_2 基传感器检测 BPA 的干扰可忽略。

MI-TiO_2 上优异的 BPA 检测性能，可以归因于其改进的导电率、优异的阳极活性、出色的分子识别能力和 TiO_2 基底锚定的丰富表面—OH 基团对于 BPA 的选择性富集。在 MI-TiO_2 的制备过程中，TiO_2 的—OH 基团与 BPA 模板之间形成几何导向的 H 键。BPA 分子插入 TiO_2 基底。热力学去除印迹 BPA，形成了充斥大量特定空腔的 TiO_2，空腔的形状、尺寸和化学功能与 BPA 互补。该无机框架分子印迹技术主要是固体基底的表面印迹，在 BPA 模板的印迹点周围形成—OH 基团。—OH 基团通过形成多个 H 键，充当主要识别组分。这导致了 MI-TiO_2 表面改善的 BPA 富集性能，BPA 浓度大大提高。另外，形状选择性和尺寸匹配也有助于 BPA 识别。

7.3.4

实际样品中双酚 A 的检测

将 MI-TiO_2/GCE 应用于各种实际样品，来评估其实用潜力[36]。在优化测试条件下，记录了没有电辅助聚集的从 0.1～1.0 V 的 DPV 曲线。实际样品中的 BPA 浓度很低，很难被 MI-TiO_2 基传感器和 HPLC 直接检测到（表 7.4、表 7.5）。因此，采取标准添加方法来考察其检测准确性和回收率[36]。每个样品经过五次平行检测，相对标准偏差（RSD）小于 5.0%。MI-TiO_2 基传感器上，所有样品加入标准 5.0～10.0 nmol·L^{-1} BPA 后，计算得到的表观回收因子在 93.2%～109.8%范围内（表 7.4、表 7.5）。该结果证明了很好的检测准确度与很弱的基质效应，这两点在实际样品分析中至关重要。而且，基于 TiO_2 的电化学方法获得的结果与标准 HPLC 结果高度一致（表 7.4、表 7.5）[23]，进一步说明了它在实际应用上的巨大潜力。

表 7.4 MI-TiO$_2$/GCE 电化学检测实际环境样品中的痕量 BPA[①]

测试样		环境样品					
		自然水	湖水	河水	污水 1	污水 2	市政污泥
测量值[①] (nmol·L^{-1})	EC	/[②]	/	/	/	/	/
	HPLC[③]	/	/	/	/	/	/
加入标准值及其测量值 (nmol·L^{-1})[②]	EC	5	5.15±0.12	4.78±0.09	5.40±0.08	4.54±0.12	5.39±0.14
		10	9.46±0.25	10.74±0.17	10.87±0.16	9.83±0.23	10.86±0.26
		15	15.29±0.41	15.60±0.27	14.52±0.324	15.27±0.27	15.83±0.32
		20	20.32±0.53	19.70±0.45	19.95±0.50	19.59±0.39	20.06±0.47
	HPLC[③]	5	8.05±0.14	7.47±0.11	6.55±0.19	6.26±0.23	5.63±0.19
		10	13.22±0.26	12.07±0.19	10.52±0.38	9.48±0.44	9.48±0.37
		15	18.10±0.47	15.06±0.32	15.17±0.44	11.95±0.69	13.51±0.54
		20	20.97±0.75	22.35±0.61	19.71±0.96	20.80±0.97	17.13±0.72
RSD (%)	EC		2.57	1.86	1.95	2.18	2.34
	HPLC[③]		2.47	1.97	3.57	4.69	3.87
回收率 (%)	EC		100.3±2.57	101.4±1.89	103.4±1.97	97.2±2.11	105.6±2.47
	HPLC[③]		129.7±3.07	120.6±2.32	109.0±3.83	101.0±4.61	95.8±3.68

测量值[①] 自然水列: EC 4.52±0.10, 9.05±0.15, 14.61±0.19, 20.18±0.35; HPLC 6.84±0.14, 9.83±0.24, 13.10±0.51, 16.32±0.68; RSD EC 1.73, HPLC 3.13; 回收率 EC 94.8±1.63, HPLC 101.0±3.00

注：① 取五次平行测量的平均值，RSD 小于 5.0%。
② 检测限之下。
③ BPA 由高效液相色谱仪 (HPLC-1100, Agilent Inc., USA) 检测，VWD 检测器波长 λ 为 254 nm，使用 Hypersil-ODS 反相柱，流速为 1.0 mL·min^{-1} 的水/甲醇 (体积比为 30∶70) 流动相。
④ 选用 0.1 mol·L^{-1} KCl 支持电解质稀释 50.0 μmol·L^{-1} 储备液。

表 7.5　MI-TiO$_2$/GCE 电化学检测实际工业样品中的痕量 BPA[①]

测试样		测量值 (nmol·L^{-1})	加入标准值 (nmol·L^{-1})[②]	测量值 (nmol·L^{-1})	RSD (%)	回收率 (%)
工业样品	奶瓶	23.3±0.78	50	72.2±2.65	3.67	97.8±3.62
	PC 水瓶	26.2±0.69	50	77.0±2.19	2.84	101.6±2.87
	饮料瓶	27.5±1.13	50	77.3±2.06	2.66	99.6±2.66
	食品袋	36.1±1.27	50	88.5±2.84	3.21	104.8±3.30
	保鲜膜	46.2±1.44	50	100.1±1.35	1.35	107.8±1.40
	餐盒	40.3±1.79	50	95.2±2.30	2.42	109.8±2.55

注：① 取五次平行测量的平均值，RSD 小于 5.0%。

② 从 20.0 μmol·L^{-1} 储备液或 0.1 mol·L^{-1} KCl 矿物盐作支持电解质稀释合适倍数。

本章小结

本章工作建立了一种分子印迹功能化的{001}-TiO$_2$ 单晶电化学检测痕量 BPA 的新方法。首先，通过调控 TiO$_2$ 的纳米片形貌和高能{001}晶面暴露，克服了其自身导电率低和催化活性弱的缺点；随后，结合无机框架分子印迹技术，显著增强了其对污染物的特定识别能力并扩大了吸附容量。由此制备的电极展现出优异的电化学传感性能，包括高灵敏性、高选择性、高稳定性和抗干扰能力。在 10.0 nmol·L^{-1}～20.0 μmol·L^{-1} BPA 浓度范围内，其 DPV 氧化峰电流与 BPA 浓度线性相关（$R^2=0.9987$），且最低检测限为 3 nmol·L^{-1}（S/N=3）。即便在多重结构类似物、腐殖酸、无机离子共存条件下，改性后的 TiO$_2$ 也能展现出良好的 BPA 特异性识别和选择性富集能力，其分析信号仍保持在 95%以上。此外，该电极也对实际环境和工业中含 BPA 的样品体现出较为出色的检测能力。本章工作证实了廉价稳定、环境友好的 TiO$_2$ 用于电化学检测环境内分泌干扰物的潜能，为新型环境检测电极的设计开发提供了借鉴思路。

参考文献

[1] Liu Q, Zhou Q F, Jiang G B. Nanomaterials for analysis and monitoring of

emerging chemical pollutants[J]. TrAC, Trends Anal. Chem., 2014, 58: 10-22.

[2] Govindhan M, Adhikari B R, Chen A C. Nanomaterials-based electrochemical detection of chemical contaminants[J]. RSC Adv., 2014, 4: 63741-63760.

[3] Chen A, Ostrom C. Palladium-based nanomaterials: synthesis and electrochemical applications[J]. Chem. Rev., 2015, 115: 11999-12044.

[4] Zhang W, Zhu S, Luque R, et al. Recent development of carbon electrode materials and their bioanalytical and environmental applications[J]. Chem. Soc. Rev., 2016, 45: 715-752.

[5] Yanez-Sedeno P, Riu J, Pingarron J M, et al. Electrochemical sensing based on carbon nanotubes[J]. TrAC, Trends Anal. Chem., 2010, 29: 939-953.

[6] Liu Y, Dong X, Chen P. Biological and chemical sensors based on graphene materials[J]. Chem. Soc. Rev., 2012, 41: 2283-2307.

[7] Wang T Y, Du K Z, Liu W L, et al. Electrochemical sensors based on molybdenum disulfide nanomaterials [J]. Electroanalysis, 2015, 27: 2091-2097.

[8] Li Y H, Liu P F, Pan L F, et al. Local atomic structure modulations activate metal oxide as electrocatalyst for hydrogen evolution in acidic water [J]. Nat. Commun., 2015, 6: 8064.

[9] Chen D, Chen C, Baiyee Z M, et al. Nonstoichiometric oxides as low-cost and highly-efficient oxygen reduction/evolution catalysts for low-temperature electrochemical devices [J]. Chem. Rev., 2015, 115: 9869-9921.

[10] Chen X, Mao S S. Titanium dioxide nanomaterials: synthesis, properties, modifications, and applications[J]. Chem. Rev., 2007, 107: 2891-2959.

[11] Kesselman J M, Weres O, Lewis N S, et al. Electrochemical production of hydroxyl radical at polycrystalline Nb-doped TiO_2 electrodes and estimation of the partitioning between hydroxyl radical and direct hole oxidation pathways[J]. J. Phys. Chem. B, 1997, 101: 2637-2643.

[12] Cho K, Qu Y, Kwon D, et al. Effects of anodic potential and chloride ion on overall reactivity in electrochemical reactors designed for solar-powered wastewater treatment[J]. Environmental science & technology, 2014, 48: 2377-2384.

[13] Kim J, Kwon D, Kim K, et al. Electrochemical production of hydrogen coupled with the oxidation of arsenite[J]. Environ. Sci. Technol., 2014, 48: 2059-2066.

[14] Yang Y, Li J, Wang H, et al. An electrocatalytic membrane reactor with self-cleaning function for industrial wastewater treatment[J]. Angew. Chem., Int. Ed., 2011, 50: 2148-2150.

[15] Yang Y, Wang H, Li J, et al. Novel functionalized nano-TiO_2 loading electrocatalytic membrane for oily wastewater treatment[J]. Environ. Sci. Technol., 2012, 46: 6815-6821.

[16] Bai J, Zhou B. Titanium dioxide nanomaterials for sensor applications[J]. Chem. Rev., 2014, 114: 10131-10176.

[17] Hanssen B L, Siraj S, Wong D K Y. Recent strategies to minimise fouling in electrochemical detection systems[J]. Rev. Anal. Chem., 2016, 35: 1-28.

[18] Hu L, Huo K, Chen R, et al. Recyclable and high-sensitivity electrochemical biosensing platform composed of carbon-doped TiO_2 nanotube arrays[J]. Anal. Chem., 2011, 83: 8138-8144.

[19] Liu L, Chen X. Titanium dioxide nanomaterials: self-structural modifications[J]. Chem. Rev., 2014, 114: 9890-9918.

[20] Liu G, Yang H G, Pan J, et al. Titanium dioxide crystals with tailored facets[J]. Chem. Rev., 2014, 114: 9559-9612.

[21] Liu S G, Yu J G, Jaroniec M. Anatase TiO_2 with dominant high-energy {001} facets: synthesis, properties, and applications[J]. Chem. Mater., 2011, 23: 4085-4093.

[22] Zhang A Y, Long L L, Liu C, et al. Electrochemical degradation of refractory pollutants using TiO_2 single crystals exposed by high-energy {001} facets[J]. Water Res., 2014, 66: 273-282.

[23] Zhou W Y, Liu J Y, Song J Y, et al. Surface-electronic-state-modulated, single-crystalline {001} TiO_2 nanosheets for sensitive electrochemical sensing of heavy-metal ions[J]. Anal. Chem., 2017, 89: 3386-3394.

[24] Kriz D, Ramström O, Mosbach K. Peer reviewed: molecular imprinting: new possibilities for sensor technology[J]. Anal. Chem., 1997, 69: 345-349.

[25] Sellergren B. Noncovalent molecular imprinting: antibody-like molecular recognition in polymeric network materials[J]. TrAC, Trends Anal.

Chem., 1997, 16: 310-320.

[26] Chen L, Xu S, Li J. Recent advances in molecular imprinting technology: current status, challenges and highlighted applications[J]. Chem. Soc. Rev., 2011, 40: 2922-2942.

[27] Jenkins A L, Yin R, Jensen J L. Molecularly imprinted polymer sensors for pesticide and insecticide detection in water[J]. Analyst, 2001, 126: 798-802.

[28] Ramstrom O, Skudar K, Haines J, et al. Food analyses using molecularly imprinted polymers[J]. J. Agric. Food Chem., 2001, 49: 2105-2114.

[29] Cai D, Ren L, Zhao H, et al. A molecular-imprint nanosensor for ultrasensitive detection of proteins[J]. Nat. Nanotechnol., 2010, 5: 597-601.

[30] Shen X, Zhu L, Wang N, et al. Molecular imprinting for removing highly toxic organic pollutants[J]. Chem. Commun., 2012, 48: 788-798.

[31] Chai S, Zhao G, Zhang Y N, et al. Selective photoelectrocatalytic degradation of recalcitrant contaminant driven by an n-P heterojunction nanoelectrode with molecular recognition ability[J]. Environ. Sci. Technol., 2012, 46: 10182-10190.

[32] Luo X, Deng F, Min L, et al. Facile one-step synthesis of inorganic-framework molecularly imprinted TiO_2/WO_3 nanocomposite and its molecular recognitive photocatalytic degradation of target contaminant[J]. Environ. Sci. Technol., 2013, 47: 7404-7412.

[33] Luo X, Zhang K, Luo J, et al. Capturing lithium from wastewater using a fixed bed packed with 3-D MnO_2 ion cages[J]. Environ. Sci. Technol., 2016, 50: 13002-13012.

[34] Ragavan K V, Rastogi N K, Thakur M S. Sensors and biosensors for analysis of bisphenol-A[J]. TrAC, Trends Anal. Chem., 2013, 52: 248-260.

[35] Ballesteros-Gomez A, Rubio S, Perez-Bendito D. Analytical methods for the determination of bisphenol A in food[J]. J. Chromatogr. A, 2009, 1216: 449-469.

[36] Hu L S, Fong C C, Zhang X M, et al. Au nanoparticles decorated TiO_2 nanotube arrays as a recyclable sensor for photoenhanced electrochemical detection of bisphenol A[J]. Environ. Sci. Technol., 2016, 50: 4430-4438.

[37] Hou C, Tang W X, Zhang C, et al. A novel and sensitive electrochemical sensor for bisphenol A determination based on carbon black supporting ferroferric oxide nanoparticles[J]. Electrochim. Acta, 2014, 144: 324-331.

[38] Wu C, Cheng Q, Li L Q, et al. Synergetic signal amplification of graphene-Fe_2O_3 hybrid and hexadecyltrimethylammonium bromide as an ultrasensitive detection platform for bisphenol A[J]. Electrochim. Acta, 2014, 115: 434-439.

[39] Hadjmohammadi M R, Saeidi I. Determination of bisphenol A in Iranian packaged milk by solid-phase extraction and HPLC[J]. Monatsh. Chem., 2010, 141: 501-506.

[40] Zhao R S, Wang X, Yuan J P. Highly sensitive determination of tetrabromobisphenol A and bisphenol A in environmental water samples by solid-phase extraction and liquid chromatography-tandem mass spectrometry [J]. J. Sep. Sci., 2010, 33: 1652-1657.

[41] Wang X, Diao C P, Zhao R S. Rapid determination of bisphenol A in drinking water using dispersive liquid-phase microextraction with in situ derivatization prior to GC-MS[J]. J. Sep. Sci., 2009, 32: 154-159.

[42] Wang X, Zeng H, Zhao L, et al. Selective determination of bisphenol A (BPA) in water by a reversible fluorescence sensor using pyrene/dimethyl β-cyclodextrin complex[J]. Anal. Chim. Acta, 2006, 556: 313-318.

[43] Wang S, Wei X, Du L, et al. Determination of bisphenol A using a flow injection inhibitory chemiluminescence method[J]. Luminescence, 2005, 20: 46-50.

[44] Vega D, Agui L, Gonzalez-Cortes A, et al. Electrochemical detection of phenolic estrogenic compounds at carbon nanotube-modified electrodes[J]. Talanta, 2007, 71: 1031-1038.

[45] Fan H, Li Y, Wu D, et al. Electrochemical bisphenol A sensor based on N-doped graphene sheets[J]. Anal. Chim. Acta, 2012, 711: 24-28.

[46] Mita D G, Attanasio A, Arduini F, et al. Enzymatic determination of BPA by means of tyrosinase immobilized on different carbon carriers[J]. Biosens. Bioelectron., 2007, 23: 60-65.

[47] Portaccio M, Di Tuoro D, Arduini F, Lepore M, Mita D G, Diano N, Mita L, Moscone D. A thionine-modified carbon paste amperometric biosensor for catechol and bisphenol A determination[J]. Biosens. Bioelectron.,

2010, 25: 2003-2008.

[48] Yin H, Zhou Y, Ai S, et al. Sensitivity and selectivity determination of BPA in real water samples using PAMAM dendrimer and CoTe quantum dots modified glassy carbon electrode[J]. J. Hazard. Mater., 2010, 174: 236-243.

[49] Yin H, Cui L, Chen Q, et al. Amperometric determination of bisphenol A in milk using PAMAM-Fe_3O_4 modified glassy carbon electrode[J]. Food Chem., 2011, 125: 1097-1103.

[50] Gao Y, Cao Y, Yang D, et al. Sensitivity and selectivity determination of bisphenol A using SWCNT-CD conjugate modified glassy carbon electrode [J]. J. Hazard. Mater., 2012, 199-200: 111-118.

[51] Mazzotta E, Malitesta C, Margapoti E. Direct electrochemical detection of bisphenol A at PEDOT-modified glassy carbon electrodes[J]. Anal. Bioanal. Chem., 2013, 405: 3587-3592.

[52] Li Y G, Gao Y, Cao Y, et al. Electrochemical sensor for bisphenol A determination based on MWCNT/melamine complex modified GCE[J]. Sens. Actuators, B, 2012, 171: 726-733.

[53] Lu S, Fei J, He Q, et al. Application of carbon glacy electrode coated with multi-wall nanotube film for determination of bisphenol A in plastic waste samples[J]. Chem. Anal., 2004, 49: 607-617.

[54] Liu X, Feng H, Liu X, et al. Electrocatalytic detection of phenolic estrogenic compounds at NiTPPS carbon nanotube composite electrodes[J]. Anal. Chim. Acta, 2011, 689: 212-218.

[55] Li J H, Kuang D Z, Feng Y L, et al. Voltammetric determination of bisphenol A in food package by a glassy carbon electrode modified with carboxylated multi-walled carbon nanotubes[J]. Microchim. Acta, 2011, 172: 379-386.

[56] Han J, Li F, Jiang L, et al. Electrochemical determination of bisphenol A using a polyacrylamide-multiwalled carbon nanotube-modified glassy carbon electrode[J]. Anal. Methods, 2015, 7: 8220-8226.

[57] Lin Y Q, Liu K Y, Liu C Y, et al. Electrochemical sensing of bisphenol A based on polyglutamic acid/amino-functionalised carbon nanotubes nanocomposite[J]. Electrochim. Acta, 2014, 133: 492-500.

[58] Li Q, Li H, Du G F, et al. Electrochemical detection of bisphenol A

mediated by $[Ru(bpy)_3]^{2+}$ on an ITO electrode[J]. J. Hazard. Mater., 2010, 180: 703-709.

[59] Huang W. Voltammetric determination of bisphenol A using a carbon paste electrode based on the enhancement effect of cetyltrimethylammonium bromide (CTAB)[J]. Bull. Korean Chem. Soc., 2005, 26: 1560-1564.

[60] Yin H S, Zhou Y L, Ai S Y. Preparation and characteristic of cobalt phthalocyanine modified carbon paste electrode for bisphenol A detection [J]. J. Electroanal. Chem., 2009, 626: 80-88.

[61] Wang F, Yang J, Wu K. Mesoporous silica-based electrochemical sensor for sensitive determination of environmental hormone bisphenol A[J]. Anal. Chim. Acta, 2009, 638: 23-28.

[62] Sun J Y, Liu Y, Lv S M, et al. An electrochemical aensor based on nitrogen-doped carbon nanofiber for bisphenol A determination [J]. Electroanalysis, 2016, 28: 439-444.

[63] Ntsendwana B, Mamba B B, Sampath S, et al. Electrochemical detection of bisphenol A using Graphene-Modified glassy carbon electrode[J]. Int. J. Electrochem. Sci., 2012, 7: 3501-3512.

[64] Zheng Z, Du Y, Wang Z, et al. Pt/graphene-CNTs nanocomposite based electrochemical sensors for the determination of endocrine disruptor bisphenol A in thermal printing papers[J]. Analyst, 2013, 138: 693-701.

[65] Huang C, Wu Y, Chen J, et al. Synthesis and electrocatalytic activity of 3Au-1Pd alloy nanoparticles/graphene composite for bisphenol A detection [J]. Electroanalysis, 2012, 24: 1416-1423.

[66] Yu C, Gou L, Zhou X, et al. Chitosan-Fe_3O_4 nanocomposite based electrochemical sensors for the determination of bisphenol A [J]. Electrochim. Acta, 2011, 56: 9056-9063.

[67] Zhang Y, Cheng Y, Zhou Y, et al. Electrochemical sensor for bisphenol A based on magnetic nanoparticles decorated reduced graphene oxide[J]. Talanta, 2013, 107: 211-218.

[68] Tu X M, Yan L S, Luo X B, et al. Electroanalysis of bisphenol A at a multiwalled carbon nanotubes-gold nanoparticles modified glassy carbon electrode[J]. Electroanalysis, 2009, 21: 2491-2494.

[69] Yin H, Cui L, Ai S, et al. Electrochemical determination of bisphenol A at Mg-Al-CO_3 layered double hydroxide modified glassy carbon electrode[J].

Electrochim. Acta, 2010, 55: 603-610.

[70] Yin H, Zhou Y, Xu J, et al. Amperometric biosensor based on tyrosinase immobilized onto multiwalled carbon nanotubes-cobalt phthalocyanine-silk fibroin film and its application to determine bisphenol A[J]. Anal. Chim. Acta, 2010, 659: 144-150.

[71] Pereira G F, Andrade L S, Rocha R C, et al. Electrochemical determination of bisphenol A using a boron-doped diamond electrode[J]. Electrochim. Acta, 2012, 82: 3-8.

[72] Ozcan A. Synergistic effect of lithium perchlorate and sodium hydroxide in the preparation of electrochemically treated pencil graphite electrodes for selective and sensitive bisphenol A detection in water samples [J]. Electroanalysis, 2014, 26: 1631-1639.

[73] Wang Y C, Cokeliler D, Gunasekaran S. Reduced graphene oxide/carbon nanotube/gold nanoparticles nanocomposite functionalized screen-printed electrode for sensitive electrochemical detection of endocrine disruptor bisphenol A[J]. Electroanalysis, 2015, 27: 2527-2536.

[74] Najafi M, Khalilzadeh M A, Karimi-Maleh H. A new strategy for determination of bisphenol A in the presence of Sudan I using a ZnO/CNTs/ionic liquid paste electrode in food samples[J]. Food Chem., 2014, 158: 125-131.

[75] Ahmed J, Rahman M M, Siddiquey I A, et al. Efficient bisphenol-A detection based on the ternary metal oxide (TMO) composite by electrochemical approaches[J]. Electrochim. Acta, 2017, 246: 597-605.

[76] Su B, Shao H, Li N, et al. A sensitive bisphenol A voltammetric sensor relying on AuPd nanoparticles/graphene composites modified glassy carbon electrode[J]. Talanta, 2017, 166: 126-132.

[77] Cosio M S, Pellicanò A, Brunetti B, et al. A simple hydroxylated multi-walled carbon nanotubes modified glassy carbon electrode for rapid amperometric detection of bisphenol A[J]. Sens. Actuators, B, 2017, 246: 673-679.